Machine Shop

AUTHORS

DONNA BATTCHER

SHELDON ERICKSON

KAREN MARTINI

CATHLEEN ROGERS

WILLIAM B. SHENNAN

ARTHUR WIEBE

EDITORS

BETTY CORDEL

SHELDON ERICKSON

JUDITH HILLEN

GRETCHEN WINKLEMAN

DAVE YOUNGS

ILLUSTRATORS

SHELDON ERICKSON
Technical Illustrator

BRENDA DAHL
Artistic Illustrator

This book contains materials developed by the AIMS Education Foundation. **AIMS** (**A**ctivities **I**ntegrating **M**athematics and **S**cience) began in 1981 with a grant from the National Science Foundation. The non-profit AIMS Education Foundation publishes hands-on instructional materials (books and the monthly AIMS Newsletter) that integrate curricular disciplines such as mathematics, science, language arts, and social studies. The Foundation sponsors a national program of professional development through which educators may gain both an understanding of the AIMS philosophy and expertise in teaching by integrated, hands-on methods.

ISBN 1-881431-39-8

Printed in the United States of America

I HEAR, AND I FORGET

I SEE, AND I REMEMBER

I DO, AND I UNDERSTAND

–Chinese Proverb

Table of Contents

To the teacher...

Machine Shop is organized with journal entries preceeding the activities which address various types of simple machines. The journal entries are not activities per se but are included to give a story line which contains content information and problem development for students.

The story line contains the activities of a youngster who calls herself Fantastic Force. She and her friend, Mucky Mass, explore a machine shop belonging to a mechanical engineer by the name of Handy MacAnnic. Handy encourages Fantastic to discover the uses of various simple machines and the forces that affect them.

The journal entries allow students to develop concepts dealing with simple machines, to see applications of the various simple machines, to clarify relevant definitions, and to employ formulae. The journal entries are written using a youngster's terminology and lingo.

The activities in *Machine Shop* can be used without the journal entries as the content information for each lesson is found in the *Background Information* of the teachers' pages.The decision of whether or not to use these journal entries lies with the teacher.

Energy and *Systems and Interactions* are themes that permeate all the investigations in this book. In order to make some of the concepts understandable to students, some units of measurements have been simplified. The intent has been to develop activities that will allow students to get at the most basic concepts and not get lost in the abstract calculations and measurements. We trust our physicist friends will understand. The editors welcome any constructive comments on lesson content, sequencing of lessons, materials recommended, or any other relevant topic.

Greedy Gravity   Fiendish Friction   Mucky Mass   Fantastic Force   Handy MacAnnic

Supply List

This list may help you in assembling the supplies for the activities you have chosen.

acetate film
aluminum foil
balances
bathroom scale
blocks of wood (scraps)
block of wood–1" x 1" x 6"
board–2" x 6" x 8'
bolts with nuts–3" long, 1/2" diameter;
3" long, 3/8" diameter)
bricks
broom handles or 1" doweling
burlap sacks
calculators
clay
colored pencils
cooking spray
corrugated cardboard
cover stock
craft knife
design toys such as Spirograph®
doweling
film canisters
galvanized wire, 14 gauge
gloves
glue
hammer
hand lenses
hexagonal pencils
hole punch
margarine tubs
markers
masking tape
masses
meter sticks–sturdy, wooden
metric tape measures
monofilament fishing line
nails
paper clips
paper cup
paraffin wax
pegboard pieces–1/4" holes, 10 cm x 25 cm
pencils
pennies (minted after 1982)
plain paper
plastic cups
plastic playing cards
pliers
polyester thread
push pins
ring stand
rubber bands
rulers–marked in millimeters
sand
sandpaper
scissors
self-stick removable notes–3" x 3"
soap bars
spring scales
steel washers
string
synthetic rope–50' x 1/4" diameter
tagboard
tape
10-speed bicycle
thread, heavy duty
thread spools—large and small
thumbtacks
toy car
vegetable oil
washers–1/4" hole, 1 1/4" diameter
watch with second hand
waxed paper

Conceptual Overview

Machine Shop is written to provide hands-on experience in the mechanics of physics for students in grades 5-9. Students are always intrigued by how machines work and are amazed how simple machines allow them to increase their force. This natural fascination draws them into the exploration and understanding of some basic concepts of physics.

The underlying theme of every activity in this book is the conservation of energy. Machines do not create energy to get work done, they only alter the factors of work, the force used, and the distance through which that force moves (Law of the Conservation of Energy). Students progressing through the activities in this book come to realize that given the right machine, they could use their own limited strength to literally move mountains. This concept of energy is coupled with the understanding that every machine involves a trade off: If a machine alters a small force to a great force, it requires that small force to go a great distance to make the great force go a small distance.

Following is a list of more specific concepts developed in this book and a visual organization of how they are related.

General Concepts of Science

- Energy cannot be created or destroyed; it can be changed from one form to another.
- Energy comes in many forms.
- Energy is used to accomplish work.
- Machines only transfer and transform energy. They cannot create or destroy it.
- Friction is a force that resists motion.
- The types of materials in contact and the amount of lubrication between the surfaces affects how much friction occurs.

General Concepts about Mechanics

- Work is done when something is moved.
- Power is a measure of how fast work is done.
- The amount of work done on an object is determined by how much force was used to move the object and the distance the object was moved.
- The amount of energy or work put into a machine will come out of the machine in some form.
- Mechanical advantage is a comparison of force produced by a machine compared to the force applied to the machine. Mechanical advantage is inversely related to the distance the produced force moves to the distance the applied force moves.
- Efficiency is the measure of how effectively a machine has transferred energy. It is the ratio of work output to total input; or the percentage of work put into a machine that is converted to useful work output.
- Friction is a major variable that reduces the efficiency of a machine.
- Friction is a cause for the difference between the expected (ideal) mechanical advantage and the measured (actual) mechanical advantage.
- Ideal mechanical advantage is the calculated value determined from theoretical understanding. It does not take into account the effect of forces such as friction.
- Actual mechanical advantage is the comparison of the actual generated force to the applied force.

Specific Concepts about Machines

- A lever transforms energy by inversely changing applied and produced force. This change is made by adjusting how torque is applied or produced by the lever.
- Torque is a rotational force generated around a pivot. It is the product of a force and the distance the force is applied from the pivot.
- Wheels and axles act as modified levers and transform energy, inversely changing applied and produced force. Wheels and axles act as separate lever-arms rotating around a common pivot.
- Gears are a form of wheel and axle.
- Pulleys transform energy by dividing the produced force over several support lines.
- Inclined planes transform energy by changing the required amount of applied force by adjusting the distance over which it must be applied.
- Screws and wedges are modified inclined planes.

ENERGY

Energy comes in many forms.

Energy is used to accomplish work.

Energy cannot be created or destroyed; it can be changed from one form to another.

The amount of energy or work put into a machine will come out of the machine in some form.

MECHANICAL ENERGY

The energy due to the position or the movement of something

MACHINES

A machine is a device for multiplying forces or changing the direction of forces.

Levers help to lift loads with less effort. The force needed for lifting changes as the distance of the effort arm changes.

Wheels and axles are turning levers.

Gears are toothed wheels.

Gears can be used to transfer forces from one part of a machine to another.

Inclined planes can be used to raise objects with less force than a straight lift.

Inclined planes are flat, sloping surfaces over which objects can be pushed or pulled to move them to a higher level.

A wedge is two inclined planes placed back to back.

A screw is an inclined plane that is wrapped around a shaft.

Pulleys can be used to increase force and/or change the direction of force.

The effort needed to raise the resistance depends on the number of lines of support the pulley system contains.

WORK

Work is done when something is moved.

The amount of work done is determined by how much force was used to move the object and how far the object was moved.

EFFICIENCY

Efficiency is the measure of how effectively a machine has transferred energy.

Efficiency is the ratio of work output to total input; work put into a machine that is converted to useful work output.

Machines lose efficiency because of friction

FRICTION

Friction is a force that resists motion.

Friction can be reduced.

POWER

Power is the rate at which work is done.

MECHANICAL ADVANTAGE

Mechanical advantage is a comparison of force produced by a machine compared to the force applied to the machine.

Ideal mechanical advantage is the calculated value determined from theoretical understanding. It does not take into account the effects of forces such as friction.

Actual mechanical advantage is the comparison of the actual generated force to applied force.

Conceptual Assessment

The following questions are presented as the type students should be able to answer after completing the investigations in this book. They do not provide an exhaustive list, but present a model of how questioning might be used to assess student understanding. The questions should be used with a picture or sample of a simple machine.

What simple machine is this?

Where is work put into (applied to) this machine?

Where does work come out of this machine?

How does this machine alter the force that is put into it (applied)?

Explain how you know if this machine will increase or decrease the force put into it.

How can you determine how much this machine will alter the force put into it?

Explain how you know if the distance the applied force moves will be greater or less than the distance the output force moves.

How does the distance and force applied to this machine relate to the distance and force generated by this machine?

How could you test to see if you were correct on your predictions about force and distance on this machine?

Explain what might cause the actual outcome to be different from your predictions?

How will friction affect this machine?

How could you make this machine more efficient?

What are some examples of other uses for this machine?

What could you change about this machine to alter the applied force?

Why would you choose to use this machine rather than to do the job without it?

What drawbacks would exist for using this machine?

Science Processing Skills

	Observing	Classifying	Measuring	Comparing	Collecting and recording data	Interpreting data	Predicting/inferring	Identifying and controlling variables	Making and testing hypotheses	Applying/general-izing/concluding
All Wound Up	X		X	X	X	X			X	X
Slip and Slide	X		X		X	X				X
Penny Weight	X		X	X	X	X				X
Clever Lever 1	X		X		X	X	X			X
Sandbagging the Seesaw	X		X		X	X	X		X	X
Clever Lever 2	X		X		X	X				X
Clever Lever 3	X		X	X	X	X				X
Push-a-Door	X		X		X	X	X		X	X
May the Force be with You	X		X	X	X	X	X	X		X
The Great Catapult Caper	X		X	X	X	X	X	X		X
One Good Turn Deserves Another	X	X			X	X	X	X		X
Gearing Up	X			X	X	X	X		X	X
Tooth to Tooth	X		X		X	X				X
Designer Gears	X				X	X	X			X
Meet the Equalizer	X			X	X	X				X
Block and Tackle	X		X	X	X	X	X			X
Pulley Power	X				X	X	X			X
Making the Grade	X		X	X	X	X			X	X
The Plane Truth	X		X		X	X	X		X	X
Wedge-Ease	X				X	X	X			X
Nuts and Bolts	X				X	X	X			X
Watt Power!	X		X		X	X	X			X

Math Content

	Using computation	Using rational numbers	Estimating	Measuring	Graphing	Averaging	Identifying/analyzing patterns	Using geometry and spatial sense	Using/applying formulae	Using calculators	Using positive and negative integers
All Wound Up		X		X	X					X	
Slip and Slide				X	X	X	X		X		
Penny Weight				X	X						
Clever Lever 1	X			X			X		X		
Sandbagging the Seesaw	X			X					X	X	
Clever Lever 2	X			X	X		X		X		
Clever Lever 3	X	X		X	X					X	
Push-a-Door		X		X						X	
May the Force be with You		X		X	X	X			X	X	
The Great Catapult Caper		X		X	X	X			X	X	
One Good Turn Deserves Another		X		X							X
Gearing Up		X		X			X				
Tooth to Tooth		X						X			
Designer Gears	X	X									
Meet the Equalizer	X			X					X	X	
Block and Tackle	X			X		X					
Pulley Power			X	X	X	X				X	
Making the Grade	X		X	X	X					X	
The Plane Truth		X		X	X						
Wedge-Ease			X	X	X			X		X	
Nuts and Bolts				X	X		X		X		
Watt Power!	X			X	X				X	X	

The Machine Shop Neighbors

Getting to Know the Cast of Characters

Handy MacAnnic

Handy is a retired engineer who spends most of his time tending his vegetable and flower garden. He is Fantastic's friendly neighbor and allows Fantastic and Mucky to turn an old shed in his backyard into their club house. The shed once served as Handy's machine shop but now it is just gathering dust. Inside are some old tools and machines like levers, wheels, pulleys, inclined planes, wedges, and screws. They have not been used in years, yet their simple qualities remain to be rediscovered.

Fantastic Force

Our heroine is small of body but strong of brain. She constantly has to help her best friend Mucky Mass out of some difficulties. With a mass of only 30 kg , she lacks the ability to generate much force. Nevertheless, she is quick, wiry, and very resourceful. She relies on the machines in the shop to help her solve the problems that Mucky gets into.

Mucky Mass

Mucky is a large, lumpy fellow with a mass of 300 kg. He's our heroine's best friend. He doesn't move very fast; in fact, sometimes it is a real chore to move at all. He is always being picked on by our villains, Fiendish and Greedy. He has a huge warm and loving heart and is trying his best at all times to do the right things. He has been friends with Fantastic for years and years.

More Machine Shop Neighbors

Getting to Know the Cast of Characters

Fiendish Friction

Fiendish is one of our villains. He is always looking for ways to ruin Mucky's life. He likes to grab at him, pull on him, and in general, make his movements as slow as can be. Fiendish is a wild little creature who has many arms which are always grabbing things and slowing them down. He has a certain laugh that he uses whenever Mucky is around and that frightens Mucky tremendously.

Greedy Gravity

Greedy is one of our villains. She tries to attract Mucky. She is not very large, but she is quite strong. She is always looking around for attractive things to collect. Greedy has been yearning for Mucky to be drawn to her for a long time and does everything she can think of to keep him close to her.

My Machine Shop Journal by Fantastic Force

August 5

Dear Journal,

I am so excited because today we just moved into a house in a new neighborhood. I now live just down the street from my best friend Mucky. Now we can have even more fun together.

I also met my neighbor. His name is Mr. MacAnnic, but he told me to call him "Handy." He was weeding his front flower bed when we started moving in. I went over to talk to him. He said he used to be a mechanical engineer and that he had designed all sorts of machines. He said he likes to talk so I was invited to come visit him anytime. After dinner, he came by our house to give us some flowers and vegetables from his garden and to welcome us to the neighborhood. He seems like a very nice man.

I noticed this old shed in Handy's backyard. It was kind of run down. The windows were all dusty so I could not see inside very well. There was a big lock on the door. Tomorrow I am going to ask Handy about it.

August 6

Dear Journal,

Today Handy showed me the shed behind his house. He used it as his machine shop when he was an engineer. It was very dusty and cluttered with all kinds of tools and machines. I told Handy that my friend Mucky and I were trying to find a place to meet and that the shop would make a neat clubhouse. Handy said we were free to use it if we cleaned it up. He told me to keep it locked up and gave me a key. Tomorrow I'll go to work on the shop.

August 9

Dear Journal,

Cleaning the shop was a lot more work than I thought, but it is now clean and Mucky and I can use it.

The shop has lots of boxes, all filled with different machines . I am going to ask Handy how to use them all.

I told Handy that I did a lot of work to get the shop clean and I thought there must be an easier way. Handy laughed and told me there is always an easier way to get work done, but it takes longer. He also said there is a harder way to do the work but it gets it done faster. I could not figure out what he meant, so he found this big book called The Mechanical Encyclopedia. Handy told me to read the section on work. Here is what it says:

> Work is the transference of energy equal to the product of the component of force that acts in the direction of the motion of the point of application of the force and the distance through which the point of application moves.

Whew!! It doesn't make much sense to me. Guess I'll ask Handy about it tomorrow...

August 10

Dear Journal,

I asked Handy about what I had read about work and he took me to the pile of dirt and junk that was outside the shop door. He said that today we had to do the work of putting this stuff into the garbage can. Handy asked if I thought I could grab it all at once and lift it into the can. I told him there was no way; it is way too heavy for me. Handy told me that that would be the hard way to get the work done, but it would be done quickly. He said that I could take the little bit I could handle and lift it into the can. I would have to lift many times to get it all in, but I would get the same amount of work done as if I had lifted the whole pile at once.

Handy explained that **work is moving something.** The short way to write this is $W = f \times d$. The W stands for the amount of work that needs to be done. The f stands for the force used to do the moving. The d stands for the distance the force moves. In my case, I do not have very much force, so I have to make many trips to lift all the junk. I, having little force and using it for a long time, did the same work as someone would who was stronger and lifted the heavy pile in one short lift.

Anyway, I did a lot of work to lift that pile of junk.

August 11

Dear Journal,

As I was headed for Mucky's today, Mr. MacAnnic reminded me that tomorrow was garbage pick-up day. I went to his backyard to move the garbage can to the front curb. I struggled for a long time and finally ended up leaning on the can in exhaustion and sweat. And, you know what, the can had not budged!

Handy came back to see how I was doing. I told him I was very tired from all the work I had done trying to move the can. He said that was kind of funny; he could not see that any work had been done. He told me that I had not moved and the can had not moved; if nothing has moved, no work has been done. Imagine that!

Handy then explained that engineers call me the effort. The can is called the resistance. No matter how much force is exerted, if the effort or resistance do not move, no work is done.

Handy then helped me. His help increased the effort force and we did the work of getting the can out to the street.

Handy said the reason I had so much trouble moving the can was because of friction. I think I better look up "friction" in the encyclopedia so I will know about it before I need it again.

August 14

Dear Journal,

Mucky and I have spent several days looking through all the boxes of "machines" in the shop. Actually, I looked through the boxes and Mucky just kind of lounged around telling stories. I was kind of disappointed in the machines in the boxes. They really were not the kinds of machines I had expected. They were only parts. Well, to try to make the best of the situation, I decided to sort all the parts.

I ended up making six piles of machine parts. There was a pile of screws, a pile of wheels, a pile of wedges, a pile of ramps, a pile of levers, and a pile of pulleys. I put each pile in a box, labeled it, and put it back out of the way.

As we were leaving the shop today to go look for adventure, we saw Handy. He asked how things were going, so I told him that I was a little disappointed in the machines because I thought they would be more than just parts. He just suggested that I read up on machines.

August 15

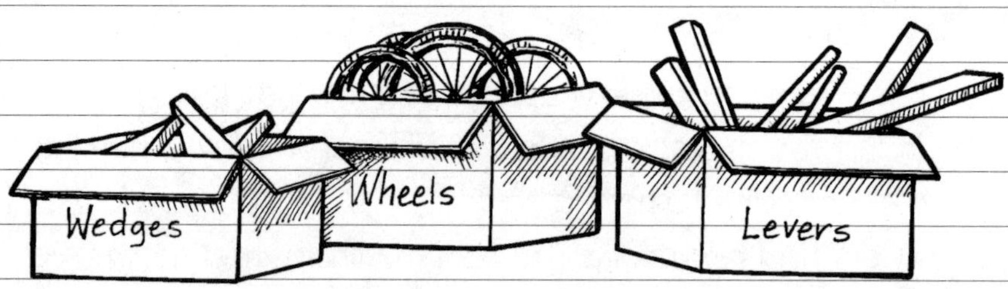

Dear Journal,

I took Handy's advice and read up on machines. Here's what I found:

> A machine is a device that transmits or modifies force or motion. A simple machine is any of six elementary mechanisms: lever, wheel and axle, pulley, screw, inclined plane, and wedge. Combinations of these simple machines form a complex machine.

Hey, great! I have sorted the parts into the right categories, but I don't understand what is meant by "elementary." Mucky and I are moving on to middle school so maybe these machines are below our intelligence. And another thing: I cannot figure how these parts can ever transmit. I thought that transmission was reserved for television and satellites. I will be checking with Handy about this.

August 16

Dear Journal,

Handy explained all this machine stuff. The six simple machines are called "elementary" because they are the building blocks of all machines. Handy says if I can understand how these six work, I can figure out how all machines work. I think it will be cool to know how my bicycle and my parents' car work! The book said that complex machines are made up lots of simple machines. If I can learn to modify and adapt all the simple machines in the boxes, I can make any machine I want. This is getting exciting. How about an automatic homework machine?

Handy explained that "transmit" means to transfer or move. Radio waves transmit, or move, television pictures from satellites to our house. Simple machines transmit, or transfer force or motion, from one place to another.

This led me to think back to our discussion on work. Handy said whatever work was done on one end of the machine would be transmitted to the other end of the machine. Then he showed me a section I had missed in the encyclopedia.

Any machine that multiplies force does so at the expense of distance. Any machine that multiplies distance does so at the expense of the force. No machine can put out more energy than is put into it. No machine can make energy; it can only transform energy from one form to another.

Handy explained that I could use a machine like a lever to lift a very heavy object with very little force. On one side of the lever, I would pull with my little force for a long distance. The lever would transform my effort and lift the heavy weight a short distance. I think I can even draw a picture to show that.

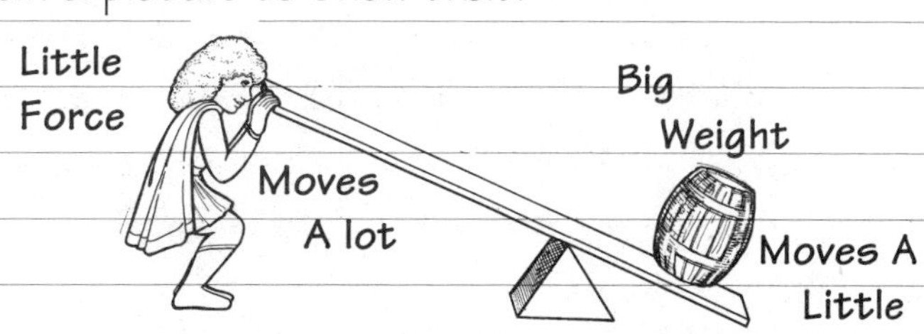

little force X **BIG DISTANCE** = **BIG FORCE** x little distance

I would have put the same amount of work into the machine as I got out of the machine. The lever did not do any work for nothing, it just changed my little force over a long distance into a lot of force over a short distance. When I figure out these machines, I will be able to transform my strength from micro-force to mega-force.

August 30

Dear Journal,

I have spent the last several days of summer getting more familiar with The Mechanical Encyclopedia. I studied up on friction.

> Friction is a force acting between two bodies in contact that are sliding or tending to slide past each other. It is caused by the irregularities between the surfaces of two objects in contact. These irregularities act as obstructions to the smooth motion of two surfaces moving past each other.

Translated, that means every material has small bumps and hollows in it. When two surfaces are rubbed together, these bumps and hollows, no matter how small, catch and stick, making it hard for the two surfaces to move over each other. Smooth surfaces have less friction than rough ones because there are fewer bumps and hollows.

The encyclopedia went on to say that there is more friction when you are trying to start something moving than when it is already moving.

Rolling something along also reduces the friction. When you roll something, it is lifted up over the bumps and hollows instead of catching on them. There is less surface area in contact. The encyclopedia also referred to lubricants.

> A lubricant is a substance, such as wax, soap, oil, or grease, used for lessening friction in the working parts of a mechanism.

It seems a lubricant kind of fills up the hollows and levels the surfaces. It keeps the surfaces from rubbing together. Lubricants are good ways to reduce friction.

It will be important to remember friction because it is always there when something moves. The machines I want to make will be full of moving parts. I need to try out some lubricants to see which will work best in my machines so I won't have so much friction. This will make my machines more efficient.

September 8

Dear Journal,

School started this week so I haven't been able to get out to the shop very much. My teacher is really nice, and we are going to do a lot of activities to discover how things work.

Dad came home with some exciting news today. He said there was going to be a tractor pull at the fairgrounds this Saturday. We made plans so the whole family can go and Dad even invited Mucky to come along if he can drag himself over to our house. I'll talk to Mucky tomorrow at school to see if he can go.

September 10

Dear Journal,

I am so excited because tomorrow is the tractor pull. Mucky gets to go! Neither of us has ever been to a "pull" so it should be fun.

I think Mucky is very tired from his first week of school. He walked slower and slower as we walked home from school. I finally had to push him up the sidewalk to his house. I'm wondering if he'll be able to make it to the tractor pull.

As I left Mucky at his house and headed for home, I heard the sinister laugh of Fiendish Friction from behind a bush. I should have known Fiendish was behind this. He always gets in the way of moving things, especially when it is Mucky that's moving.

On my way home I remembered what I'd read about lubricants. The book said that when friction got in the way, lubricants made things easier to move. I decided to have a look in the shop for some lubricant that might help Mucky. There was a whole box filled with all sorts of oils, greases, and gunk. Luckily, there's a lot of this stuff because we're going to need it for Mucky if he has trouble moving.

* * * * *

Editor's Note: Do the activities *All Wound Up* and *Slip and Slide*. Apply what you have learned from Fantastic Force's journal. Maybe you will be able to figure out how Fantastic plans to help Mucky with his movement problem.

September 11

Dear Journal,

Well, things did not go as planned today. Mucky didn't show up on time to go to the tractor pull. I raced to his house with the lubricants and found, as expected, he was having trouble moving. He had managed to get dressed and out the door, but he could not make it down the sidewalk.

I moved into action. I slimed the sidewalk with a layer of grease. Then I doused Mucky with several cans of oil. Before I realized it, Mucky was moving down the sidewalk at breakneck speed. It was going great until Mucky shot into the street. Unfortunately, a car was coming through the intersection at the same time. It stopped, but not quite in time for Mucky. The only race Mucky saw was a quick one to the hospital. Luckily, he only got a couple of scrapes and a broken ankle. He now has a cast on his leg as evidence of his speediness.

When I told Handy what had happened, he reminded me that Fiendish Friction's middle name is Friendly. Handy told me that friction can be very useful and that without friction, we could not walk because our feet would slip. We couldn't even hold a pen or pencil. Automobile brakes use friction to slow down a moving car. Friction helps hold nails and screws in wood. A certain amount of friction would have helped Mucky have a safe trip down the sidewalk. Handy reminded me that I need to take time to carefully formulate my plans and then he left me with this saying, "Too much of a good thing can be dangerous!"

All Wound Up

Topic Area
Friction

Introductory Statement
In this investigation, students will create a thread-spool tractor. They will examine the role of friction in the movement of this tractor and make modifications to it that will make it travel farther.

Math
Using rational numbers
- ratios
- proportions
- percents

Using calculators
Measuring
- length

Graphing

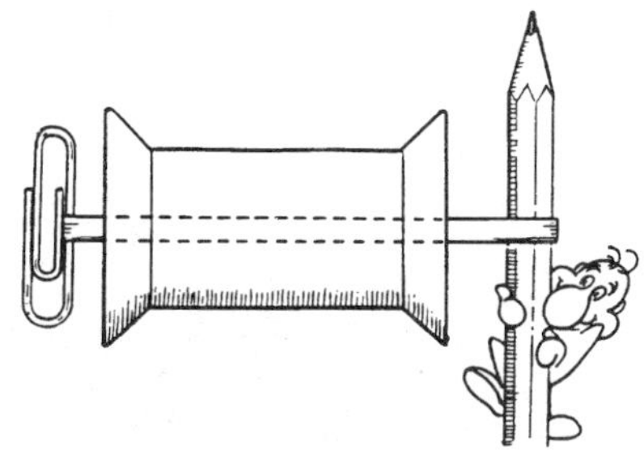

Science
Physical science
- Simple machines

Math/Science Process Skills
Observing
Making and testing hypotheses
Collecting and recording data
Comparing and contrasting
Interpreting data
Applying and generalizing

Materials
For each group:
- large plastic thread spool
- 2 paper clips
- rubber bands, a variety of sizes (See *Management*)
- pencil or drinking straw
- cooking spray or vegetable oil (lubricant)
- sheet of acetate film (overhead transparency)
- steel washer
- scissors
- hole punch

Key Question
How can you make a spool tractor that will go the farthest distance?

Background Information
The students will be constructing a spool tractor that will utilize a thread spool as a wheel, a pencil as a lever, and a rubber band as a power source. The rubber band extends through the hole in the spool. One end of the rubber band is secured to the spool with a paper clip. The other end of the rubber band is attached to a pencil or straw (see diagram on student's page). The rubber band is wound by turning the pencil. When the spool tractor is placed on the ground and released, the rubber band begins to unwind. The leverage exerted by the pencil is too great for the force of the wound rubber band to overcome. Instead, the rubber band releases its energy on the paper clip end. Because the paper clip is attached to the spool, the spool turns simultaneously with the paper clip as the rubber band unwinds. If the rubber band does not have enough force to overcome leverage of the pencil or the inertia of the spool, it is not able to unwind and the spool tractor will not move.

After building and running this basic tractor, a discussion should be held on the role of friction in the mechanics of the spool tractors.

- Friction produced by the pencil's contact with the spool hinders the movement of the spool to turn. The tighter the rubber band is wound, the tighter the pencil is pulled against the spool. This source of friction can be reduced by modifying the surface (inserting a washer or piece of acetate film), applying lubricants, or adjusting the tightness of the rubber band.
- The wound rubber band within the spool can be another source of unwanted friction. If the rubber band is too large, or wound too tightly, it will rub against the sides of the spool's hole. This friction may be enough to stop the unwinding of the rubber band. Adjustments to the rubber band and/or the number of winds would solve this problem.
- Friction plays a positive role in the contact of the spool to the rolling surface (the floor). Plastic spools tend to be slippery and if the spool has too much torque, it will slip and spin rather than move the spool forward.

A great many variables affect the efficiency of the spool tractors. After trying the basic tractor, students need to be given time to explore and test modifications. Students should have access to a variety of sizes and lengths of rubber bands so they can explore different power sources and tires. The acetate film and steel washers provide a means of making the spool's surface smoother. To test the acetate film, cut a disc the size of the spool's end and punch a hole in the center. Thread it onto the rubber band so it is positioned between the spool and the straw. Adding lubricant to the disc reduces the friction even more and may increase the distance the spool tractors will go.

Students will judge their improvement by comparing the distance of the improved tractor to the distance of the original basic tractor.

$$\frac{\text{Improved distance}}{\text{Basic distance}} = \text{Improvement ratio}$$

$$\text{Improvement ratio} \times 100 = \text{Percentage of improvement}$$

If students get 100 percent, they have not improved their tractor. The amount greater than 100 percent is the amount of improvement they have gained. The students could strive for attaining the greater distances and/or greater percentages of improvement.

This activity is written to emphasize the role of friction. It also provides an opportunity for a discussion of other aspects involved in the mechanics of the tractor. The teacher can decide if it is appropriate to consider the roles of the following:

- The elasticity of the rubber band is very influential in the tractor's movement. When students try different lengths and thicknesses of rubber bands, they are testing the effects of elasticity.
- The leverage of the pencil is an important aspect of the tractor. The effect of this leverage can be explored by changing the length of the pencil. If the pencil is shortened, the torque of the pencil becomes low enough that the spool is lifted on the pencil's side while the paper clip side remains in contact with the floor. As the spool spins, the paper clip side moves forward while the pencil side does not. Thus, the tractor responds by spinning in a circle.
- The tractor provides a good example of the third law of motion (for every action there is an opposite and equal reaction). As the rubber band unwinds clockwise, the spool moves forward with a counter-clockwise rotation.
- The transformation of energy can also be discussed with the tractor. The kinetic energy (energy of motion) of the student winds the rubber band. The rubber band stores this energy as potential energy. As the rubber band unwinds, the potential energy is converted to kinetic energy as it moves the tractor.

Management

1. Collect the materials listed. Gather a variety of rubber bands with some being close in length to the height of the spool.
2. This activity is best done in groups of two.
3. Allow a minimum of two forty-five minute class periods for this activity. The first period will be used to construct, test, and discuss the basic tractor. The second period is for the modification of the tractors. Time outside of class may be given for the second part.
4. Be certain to construct a few tractors before assigning this activity to students. Find the rubber band length that works best for various spool sizes. Generally, the thinner rubber bands work best.
5. Determine the number of rubber band winds possible before the rubber bands break. This will give a point of reference for warning students not to exceed this number of winds.
6. If students are to make between group comparisons and contrasts, rules of "fairness"–the control and manipulation of variables–need to be examined and established.
7. The distance for all tractor runs is the straight line distance between the starting and stopping points. The course the tractor follows is ignored. However, students will discover that it is to their benefit to get their tractors to go as straight as possible.

Procedure

Part 1:

1. Discuss the *Key Question.*
2. Have students pair up and distribute the materials and instructions of building the basic tractor.
3. Direct students to name their tractors and construct them according to the instructions.
4. Have students wind the rubber bands a number of times–that number being less than the previously determined breakage point number–and release their tractors.
5. Instruct students on the rules of measuring the distance of the run. Ask them to measure and record their findings on the chart.
6. Allow students the time to more carefully observe their tractors, determining how they work and where friction is involved.

Part 2:

1. Discuss with students where friction is involved in the tractors. Have them determine where friction has a positive and negative effect on the tractor's operation. Discuss how they might modify their tractors to alleviate unwanted friction or to take advantage of wanted friction.
2. Allow time for students to explore and try various modifications of their basic tractors. Materials for modification need to be made available to the students. Ask students to continue modifying their tractors until they are satisfied they have constructed the best possible tractor or until the allotted time has run out.
3. Using their modified tractors, have students wind their rubber bands the recommended number of times and release the tractor.
4. Instruct the students to measure and record the distances their modified tractors travel.
5. Ask students to share their distances with the rest of the class.
6. Have students complete the chart doing the necessary calculations.
7. Direct the students to make bar graphs of the data.
8. Give students the opportunity for observing, comparing, contrasting, and discussing the various tractors and their responses to the modifications. Discuss possible reasons for the results.

Discussion

1. Did your tractor improve with modification? If so, how much?
2. If you improved 200%, how much higher would the improved-run bar be on your graph than the basic-run bar? [twice as high]
3. What modifications did you make to your basic tractor? Why did you make these modifications? How successful were the modifications?
4. What tractors showed the most improvement? What modifications were made on these tractors?

5. Which tractors went the greatest distances? Which tractors went the least distances?
6. What similarities do you see in the tractors that went the greatest distances?...the least distances?
7. What differences do you see between the tractors that went the greatest distances and those that went the least distances?
8. How might these differences affect how far the tractors would go?
9. What other observations did you make in variables that might affect how the tractors ran?
10. If you were to modify your tractor again, how would you modify it?

Extensions

1. This same activity could be done using different types and sizes of spools, and/or different road surfaces–carpet, tile, concrete, asphalt.
2. Have students time their tractors and measure the distances the tractors go in that time. Using these measurements, the average speeds of their tractors can be determined. Have them compare and contrast these measurements to previously gathered data to see how the speeds of the tractors relate to the distances they were able to travel. Are the tractors that go the fastest also the tractors that go the greatest distances?
3. Have students research the application of friction in vehicles. How do tires differ? [tracks, treads, tractor tires, slick racing tires] How is friction reduced? [bearings, grease, oil, oil and gas additives]

Curriculum Correlation

Language Arts:

Create a crossword puzzle using terms from this activity: friction, machine, lever, potential and kinetic energy, revolution, graph, data, and so on.

Major Conceptual Components

- Energy is used to accomplish work.
- Work is done when something is moved.
- Friction is a force that resists motion.
- Friction can be reduced.

Tractor Assembly

1. Attach a rubber band to a paper clip.

2. Thread the rubber band through the hole in the spool. You may need to use another paper clip as a needle.

3. Place the pencil through the loop of the rubber band that is sticking out of the spool.

4. Attach the paper clip to one of the struts on the end of the spool opposite the pencil.

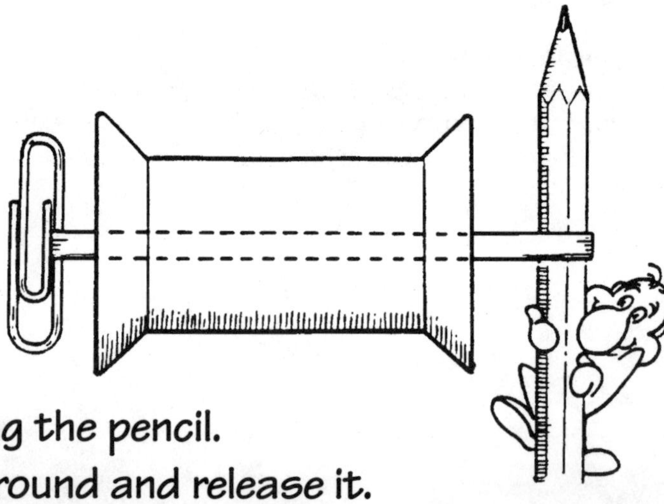

5. Wind the rubber band by turning the pencil.
6. Set the spool tractor on the ground and release it.
7. Measure how far it is from your tractor's starting to ending position.
8. Name your tractor and record on the chart its name and the distance this basic tractor design traveled.

All Wound Up

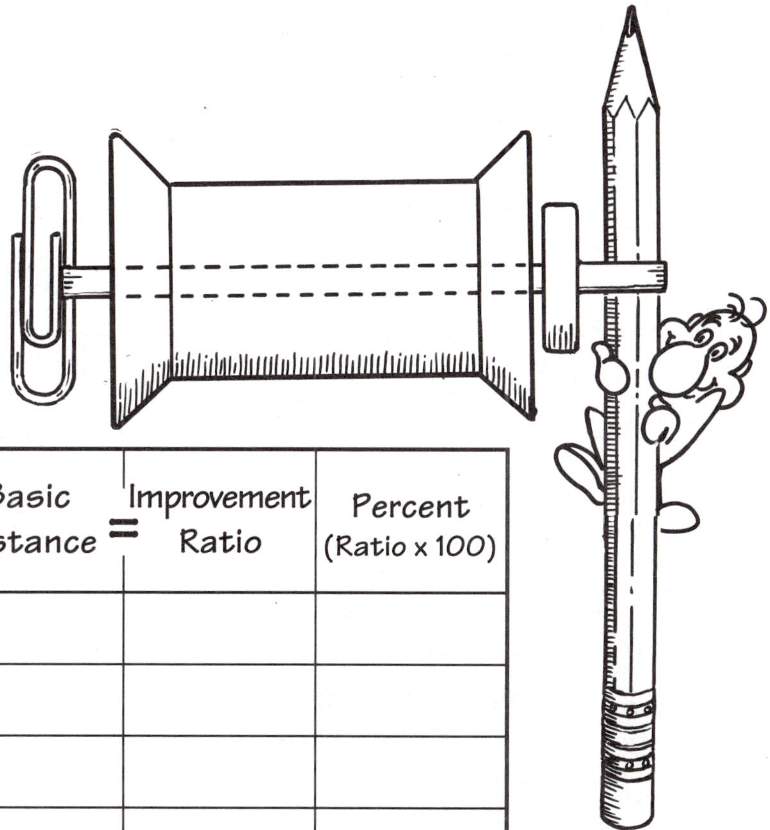

Group	Tractor Name	Improved Distance ÷	Basic Distance =	Improvement Ratio	Percent (Ratio x 100)
A					
B					
C					
D					
E					
F					
G					
H					
I					
J					
K					
L					
M					
N					
O					
P					
Class Total					
Class Average					

All Wound Up

Distance from Start to Finish

Group																Percentage of Improvement
A	Basic Tractor															
	Improved Tractor															
B	Basic Tractor															
	Improved Tractor															
C	Basic Tractor															
	Improved Tractor															
D	Basic Tractor															
	Improved Tractor															
E	Basic Tractor															
	Improved Tractor															
F	Basic Tractor															
	Improved Tractor															
G	Basic Tractor															
	Improved Tractor															
H	Basic Tractor															
	Improved Tractor															
I	Basic Tractor															
	Improved Tractor															
J	Basic Tractor															
	Improved Tractor															
K	Basic Tractor															
	Improved Tractor															
L	Basic Tractor															
	Improved Tractor															
M	Basic Tractor															
	Improved Tractor															
N	Basic Tractor															
	Improved Tractor															
O	Basic Tractor															
	Improved Tractor															
P	Basic Tractor															
	Improved Tractor															

Slip and Slide

Topic Area
Inclined planes, forces

Introductory Statement
Students will measure the degree of slope required to overcome friction on different surfaces. They will then apply a lubricant to the surfaces to test its effect.

Math
Measuring
  angles
Averaging
Graphing

Science
Physical science
  Simple machines

Math/Science Process Skills
Observing
Collecting and recording data
Interpreting data
Drawing conclusions
Applying

Materials
Per group:
- scissors
- glue
- tape
- flat 30 cm ruler
- plastic 35 mm film canister
- 10 pennies
- vegetable oil
- 2.5 cm X 10 cm strips of each of the following:
  - plain paper
  - overhead acetate film
  - aluminum foil
  - fine sandpaper
- optional: hand lenses

Key Question
What type of surface is best for making a slide?

Background Information
Friction is the force that acts to resist motion when surfaces are in contact with each other. It is caused by the irregularities in the surfaces of materials. Every surface has bumps and hollows so that when two surfaces are rubbed together, the bumps and hollows catch and stick to resist the movement of the surfaces over each other. Friction makes it hard to push one material across another.

The amount of friction between two objects is dependent upon the types of surfaces involved. Smooth surfaces produce less friction than rough surfaces. More force is needed to start an object sliding than to keep it sliding because starting friction is greater than sliding friction. And likewise, less force is needed to roll objects than to slide them because rolling friction is less than sliding friction. In rolling friction, the roller or wheel has less surface area in contact with another surface to catch in the bumps and hollows. Making the surfaces smoother by leveling out the bumps and hollows reduces the friction between them. Slippery materials called lubricants (in this activity, vegetable oil) reduce friction. The lubricant is placed between the two surfaces, filling in and leveling out the bumps and hollows.

As the slope of an incline becomes greater, the tendency of an object to move becomes greater. The angle of an incline can be used as a measure of an object's tendency to move. In this activity, the amount of friction between two objects will be determined by measuring the angle of the incline required to start the object moving past each other. The steeper the slope (the greater the angle), the greater the friction.

Management
1. This activity is best done in groups of four: one student to raise the ruler, one to measure the angle, the third to change the surfaces, and the last to record the results.
2. Allow one class period (45 minutes) for this activity.
3. If possible, copy the protractor and stand page onto cover stock. This stronger weight paper will stand up better to student use.
4. Smooth plastic is a relatively slippery surface. Film canisters are used to take advantage of this characteristic. By placing the ten pennies inside, the canister will provide enough weight to achieve adequate results. If film canisters cannot be found, other objects may be substituted, but the teacher should try out the substitution before class to judge its effectiveness.
5. A thick coat of vegetable oil is needed to act as a lubricant. As this tends to be quite messy, it is advisable to protect the desk or table surface with several layers of newspaper.

Procedure
1. Discuss the *Key Question.*
2. Distribute materials and instruct students to cut out, fold, and glue the protractor and stand as explained on the student page.
3. Have students tape the stand to the table so the protractor is in a vertical position.
4. Using tape as a hinge, have students tape the ruler to the mark on the protractor stand. The smoothest side of the ruler should be face up.
5. If you are using the optional hand lenses, invite students to observe the various surface textures of the paper, acetate film, aluminum foil, and sand paper.

6. Have students tape the paper strip to the top side of the ruler.
7. Direct students to place the bottom of the film canister which is filled with the ten pennies on top of the strip of paper.
8. Instruct one student to slowly raise the free end of the ruler until the canister begins to move.
9. Another student measures and records the angle at which the canister starts to slide.
10. The ruler is again placed flat on the table with the film canister in its original position and steps eight through ten are repeated two more times with the results for each trial being recorded.
11. The angles are totaled, averaged and recorded on the chart.
12. Steps 6 through 11 are repeated for the overhead acetate film, the aluminum foil, and the sandpaper.
13. Heavily lubricate each strip of material and the bottom of the canister with the vegetable oil.
14. Repeat steps 6 through 11 with the lubricant. Record the measurements on the second chart.
15. Have students make a bar graph of their data.

Discussion

1. What material allowed the canister to slip at the lowest angle? [Answers may vary. Plastics tend to be slippery so the acetate is usually the lowest. Sandpaper is very rough and tends to be the highest]
2. What material caused the least friction? How did you determine this?
3. How did the lubricant affect friction on each of the four materials? What problems were created when applying the lubricant? Could these problems be overcome? [Paper tends to absorb oil and often becomes tacky. If enough lubricant is used, the object will "float" on the lubricant. The sandpaper is so rough that the lubricant has little effect]
4. How does the graph display this information?
5. If you wanted to make the best slide, how would you make it?

Extensions

1. Discuss other variables that might affect friction. [weight of object, other contact surfaces, other lubricants] Have students devise their own experiments and test these variables.
2. Have students study the soles of shoes for differences in construction and materials and how the construction and materials affect the amount of friction they provide. Discuss why the soles are different for such sports as basketball and bowling. Have students devise a test to rank shoes by amount of friction.
3. Discuss with students the positive and negative aspects of friction. Have them find as many examples of each as they can.

Major Conceptual Components

- Energy is used to accomplish work.
- Work is done when something is moved.
- Friction is a force that resists motion.
- Friction can be reduced.

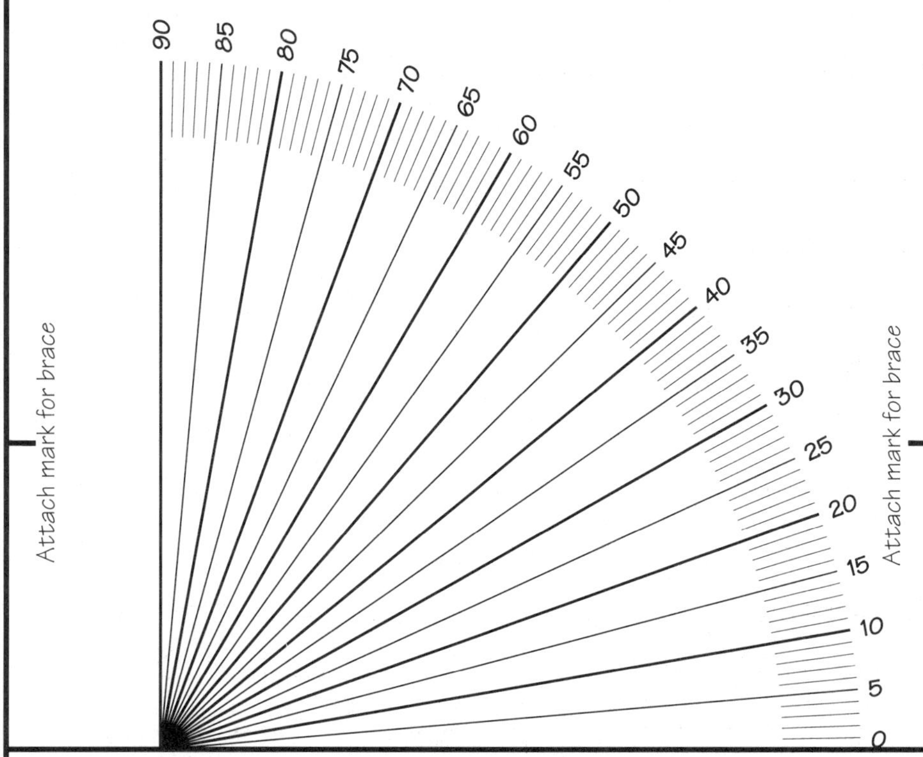

Fold along line above so protractor is vertical and this part is face up on table.

After making the protractor unit and taping it to the table, attach ruler with tape so it hinges on the line to the left and covers this writing.

Fold along line above so that this side faces the table and serves as the base.

Fold along the line above so that this side makes a diagonal brace to hold the protractor in a vertical position.

Fold along the line above and use this as a tab to attach the brace to the protractor. Make sure the protractor will stand vertically before attaching.

SLIP AND Slide

Cut out the rectangle to the left along the exterior bold line, then follow the instructions.

SLIP AND Slide

What surface makes the best slide?

Without Lubricant

	Degrees of Incline			
	Paper	Plastic	Aluminum Foil	Sand-paper
Trial 1				
Trial 2				
Trial 3				
Total				
Average				

With Lubricant

	Degrees of Incline			
	Paper	Plastic	Aluminum Foil	Sand-paper
Trial 1				
Trial 2				
Trial 3				
Total				
Average				

Average Degrees of Incline for Different Surfaces With and Without Lubricant

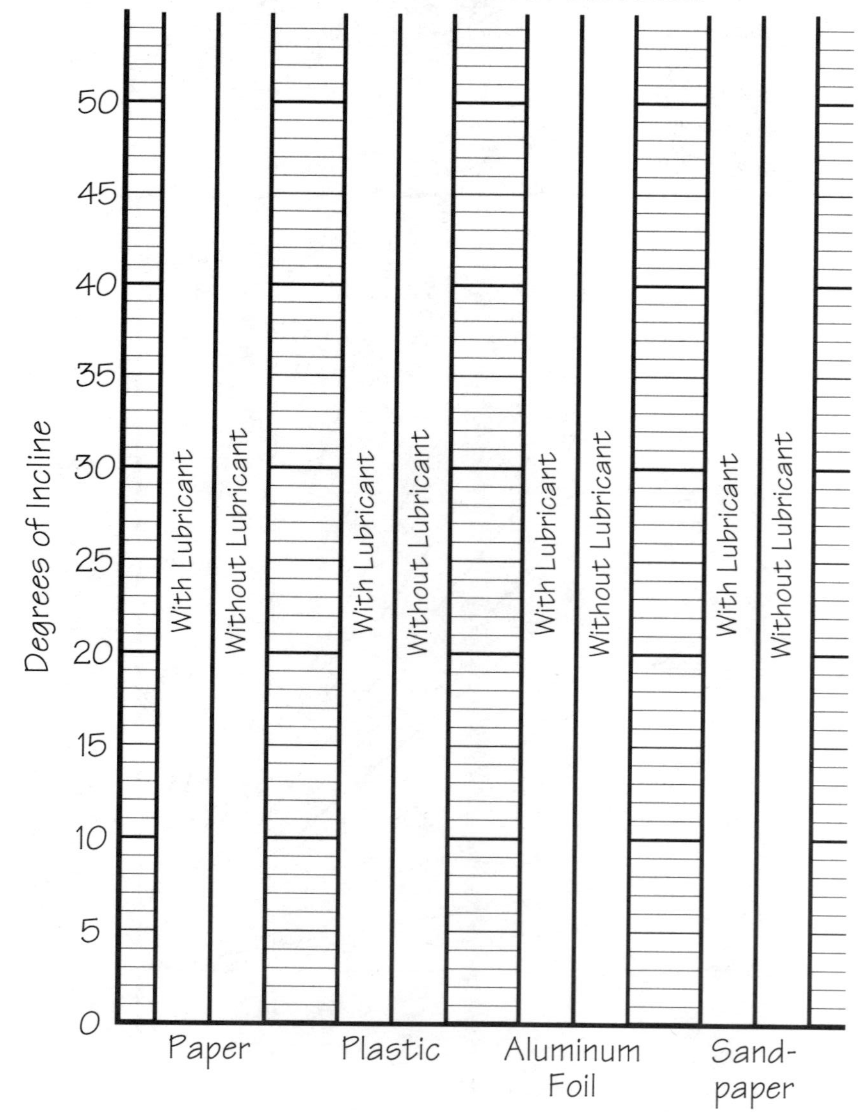

September 18

Dear Journal,

Well it's been a week since Mucky's fateful accident. He has a cast on his leg so he is not able to get around so well. Since Mucky cannot do much but sit, we decided to start playing with the machines in the shop.

The encyclopedia says that the lever is the simplest machine, so I got the box out with all the levers. Mucky and I emptied the box out on the floor and started to look at them.

This is how the encyclopedia defines levers.

> A lever is a bar or rigid body used to lift weights. The bar pivots around a fixed point called a fulcrum. A lever operates on the principle that force applied at one point lifts a resistive weight or force at a second point, by rotating it about the fulcrum.

We sorted out the parts into piles on the floor. There were lots of strong bars. Some had holes in them while others did not. There were different types of fulcrums too. Some were posts that had rods that stuck out parallel to the ground. Others were just triangular wedges.

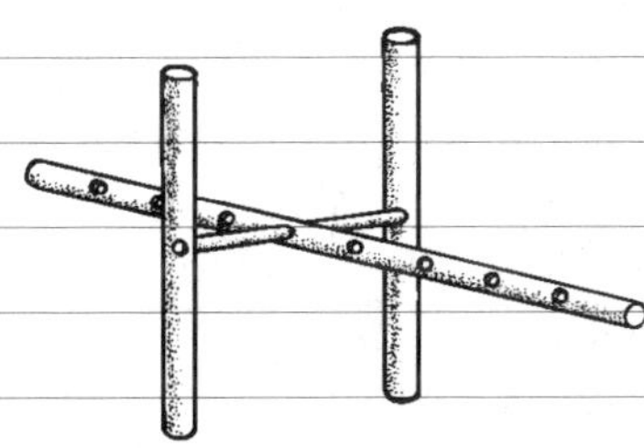

Mucky and I decided there were two ways to put the levers togethers. One way was for the rods on the post to be placed through the holes in the bars. These formed levers that would spin around the rods. We couldn't figure out why the bars had so many holes. Which hole were you suppose to use to make a lever?

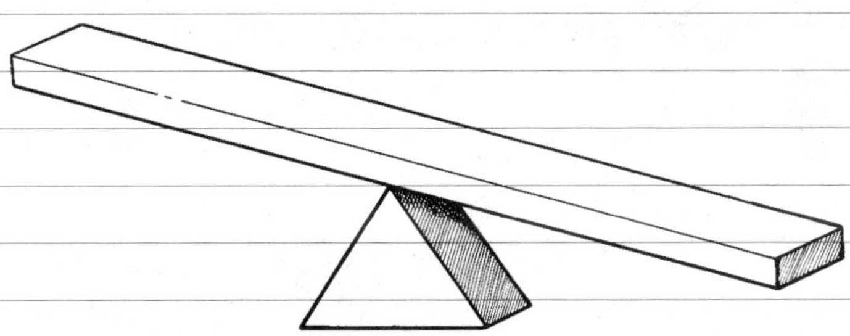

The other way was for the bars without holes to be set on the wedges. These levers would rock back and forth like a seesaw. The bars without holes could be set anywhere along the wedge. How do you decide where to place the wedge along this bar?

I guess I just need to ask Handy about where to put the bar in a lever.

September 25

Dear Journal,

Today was a wonderful fall Saturday. It was just too nice a day to sit in the machine shop, so before lunch I planned to get together with the rest of the neighborhood for a game of football.

Since I was up early, I went over to see Handy and ask him how the levers were supposed to be assembled. He was in his garage getting some tools. He said he had to get to work pulling out a tree stump, but told me to come along and he would explain how to assemble levers while he worked.

Handy had dug all around the stump and had chopped at most of the roots. The stump was still in the ground and had to be pulled out. Handy jammed one end of a heavy steel bar under the stump. He put a rock under the bar as close to the stump as he could. Handy grabbed hold of the bar down near the rock and pushed with all his might on the bar but he could not budge the stump. He asked if I would help him. I grabbed ahold of the end of the bar which was a long way from the rock. I pulled the bar down and, believe it or not, the stump went up. It was a piece of cake! Handy said that I was a first-class lever puller. He explained that he had made a first-class lever. The resistance, the thing you are trying to move, is the stump. It is at one end of the lever. The fulcrum, the place where the lever rotates, is the rock. It is somewhere in the middle of the lever. The effort, the force doing the pushing, is Handy or me. The effort is on the side of the fulcrum opposite the resistance.

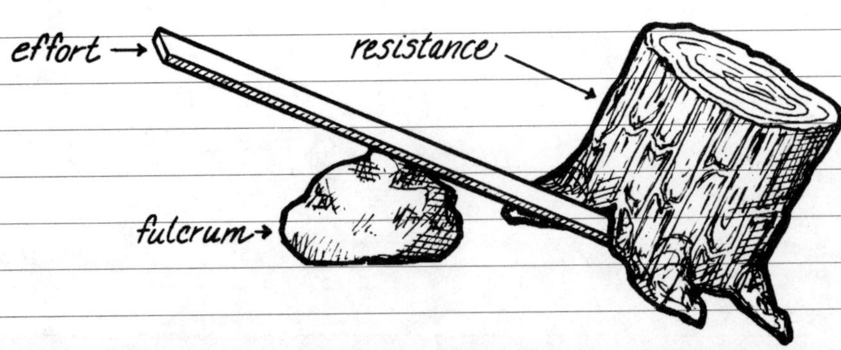

When the stump was out, Handy and I struggled to get it up into a wheel barrow. Once we got it in, I was surprised at how easy it was to move it out to the dump pile. Handy explained it was easy to move the stump because the wheel barrow is a second-class lever. In a second-class lever, the resistance (the stump) and the effort (me) are on the same side of the fulcrum. The wheel is the fulcrum because the forces rotate around it.

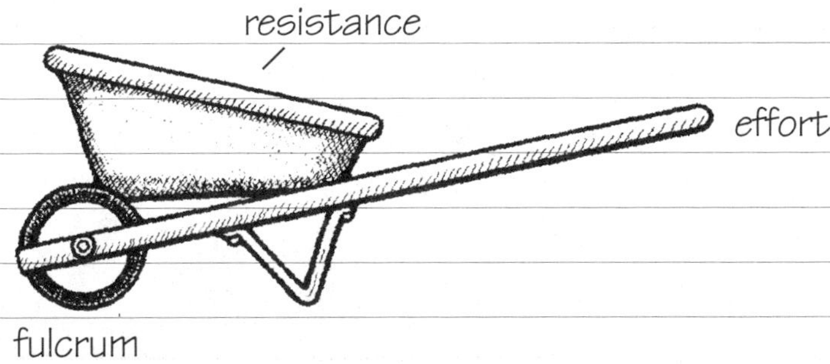

Handy was getting tired so he asked if I would sweep the sidewalk near where the stump had been. As I swept, he told me I was working a third-class lever which was the broom. The fulcrum was my hand at the top of the broom handle. The effort was my other hand, pulling the handle. The resistance was the dirt I was sweeping. Like a second-class lever, a third-class lever has the resistance and effort on the same side of the fulcrum. But in a third-class lever, the effort is closer to the fulcrum than the resistance is.

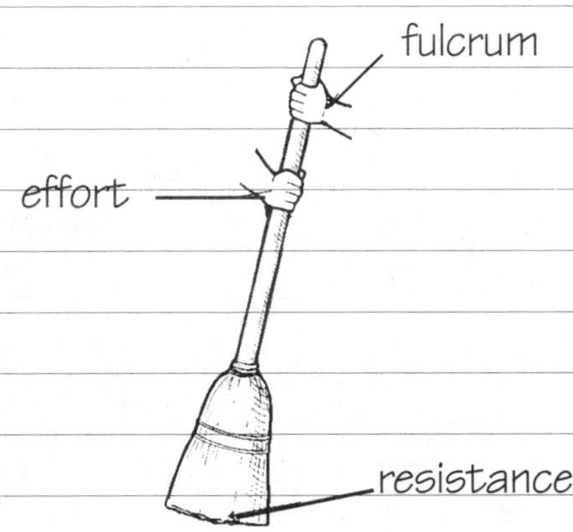

When I had finished sweeping, I realized I was late for the game. I raced off but still only got in on the tail end. I walked home with Mucky who had been watching the game and explained what Handy had told me about levers. It was then that I realized that I knew three different ways to assemble levers. That still doesn't explain why there are so many holes in some of the bars or how to decide where to put the fulcrum. I guess that means I have something more to ask Handy. I hope he doesn't get tired of all these questions.

October 2

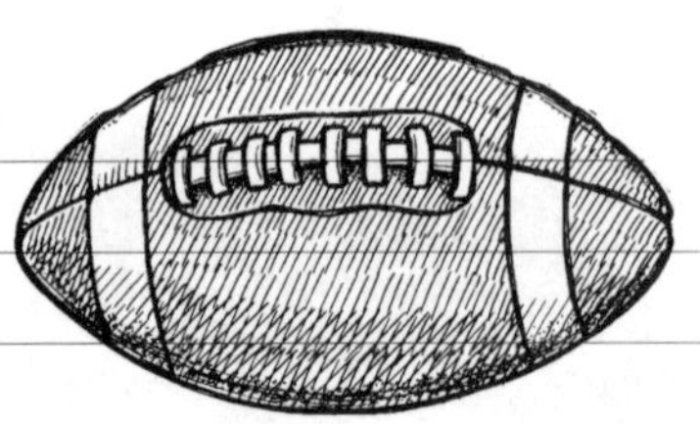

Dear Journal,

Another football game was scheduled for this morning. I decided to wait until afternoon to ask Handy my questions. Of course, he was off on some errand so I had to do the research on my own. I continued reading about levers in the encyclopedia.

> The length of lever arms required in a system is calculated by determining the torque generated by each arm. Rotation will take place in the direction of the arm with greater torque. Torque is calculated by finding the product of the force applied and the distance of that force from the fulcrum.
>
> Torque Equation: Torque = force x distance

Now this stuff is getting into algebra. I need to wait to talk to Handy.

October 9

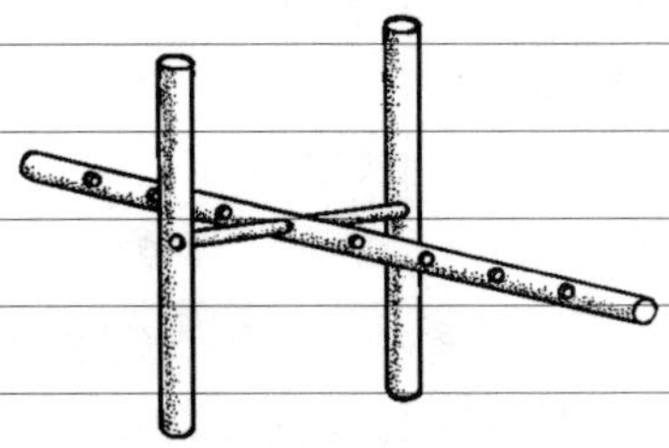

Dear Journal,

When I asked Handy about torque and if it had any thing to do with the holes in all the bars, he told me to remember about taking out the stump. He pushed with all his weight on the bar and could not move the stump. He had pushed with the force of all his weight. Handy explained that we use newtons (abbreviated N) to measure force. A mass of 1 kg pushes down with a force of 9.8 N. So each kilogram is about the same as 10 newtons. Handy has a mass of 60 kg. The most force he could push down with would be 600 N. With 600 N, he could not move the stump when he was only 1.0 m from the rock. But my mass of 30 kg made 300 N and I easily moved the stump. The difference was that I was 3.0 m from the fulcrum. Handy showed me how to calculate the amount of torque we both generated.

Handy: 600 N x 1.0 m = 600 units of torque (newton-meters)
Me: 300 N x 3.0 m = 900 units of torque

I generated more force then Handy because I was farther from the fulcrum than Handy. That's dandy, isn't it? Oops, I'd better quit with the rhymes or I won't get finished.

Handy went on to explain that with machines, strengths are compared by seeing how much force comes out of the machine compared to how much force was put into it. This is called the mechanical advantage. Handy figures the tree stump has a mass of 90 kg, making a force of 900 N. With my weight I can generate only 300 N of force. The lever changed my 300 N into 900 N. To calculate mechanical advantage I just make the ratio:

resistance force / effort force = mechanical advantage (M.A.)

900 N / 300 N = 3

My force was made 3 times bigger by using the machine. I'm impressed!

I can change the torque and mechanical advantage by moving the fulcrum. With the levers, I can slide the wedges, or put the rods through the holes to give me the torque I will need.

October 14

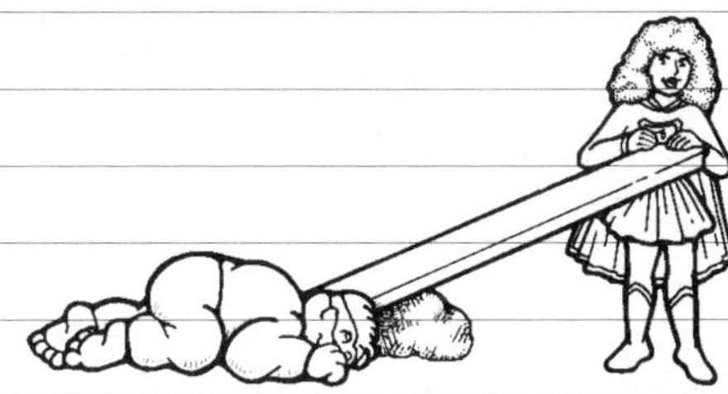

Dear Journal,

We decided to play football at the park after school today. It was definitely fall as it is getting dark so much earlier, and there was that chill in the air. I was walking by the school yard when I heard a wimpy cry from the playground. It was Mucky. He had managed to drag himself, with the cast, to the top of the slide. Exhausted, he had only enough strength to lie on his stomach and go head-first down the slide. When I got there, he lay on his stomach at the bottom of the slide with no strength to pull himself up. He was moaning that he was so tired and that he just knew he would freeze to death in the dark if I could not help him to stand up.

I was glad I had talked to Handy. First, I tried dragging Mucky, but there was too much friction. I remembered that one way to reduce friction was to roll something. So, I rolled Mucky over to a rock near the seesaw. I took the board off the seesaw. I stuck one end under Mucky and put the board over the rock. Then I climbed onto the board and started crawling out toward the other end. As I got almost to the end, the seesaw started to move. Slowly Mucky was lifted off the ground. As his legs hung down, he was able to stand up and get off. Well, the minute he stood up, I dropped. OUCH! Science can be painful.

As we left the playground, I turned and saw two figures in the shadows. They were muttering. I overheard one say, "Drats! He was pulled away from me again!" It was then that I realized I had torn Mucky away from the forces of Greedy Gravity.

I was so excited about my new-found strength that I came home and did these calculations with the estimated lengths:

Mucky = 300 kg ≈ 3000 N   Fantastic = 30 kg ≈ 300 N

3000 N (resistance) / 300 N (effort) = 10 (M.A.)

Mucky's Torque		My Torque
3000 N x 0.4 m	<	300 N x 4.5 m
1,200 Nm	<	1,350 Nm

This explains why the seesaw tipped my direction. This machine stuff could be real useful!

Penny Weight

Topic Area
Levers and leverage

Introductory Statement
Using a meter stick and a pencil, students will construct a lever; they will gain an intuitive knowledge of leverage by adding an increasing number of pennies to one end and then by observing the necessary change in the length of the portion of meter stick needed to balance the end with pennies.

Math
Measuring
  length
Graphing

Science
Physical science
  Simple machines

Math/Science Process Skills
Observing
Collecting and recording data
Comparing and contrasting
Interpreting data
Applying

Materials
For each group:
  1 rigid meter stick
  tape, string, paper clip
  1 hexagonal pencil
  paper or plastic cup
  50 pennies, minted in or after 1982

Key Question
How must the location of the fulcrum change to keep a system in balance when weight is added to one end?

Background Information

> A force is a push or pull. Forces tend to make moving things stop and stationary things begin to move. Forces can act against each other on the same object to produce a net force of zero.

> Simple machines are devices with few or no moving parts. They may change the direction and/or the size of forces applied to them. A list of simple machines usually includes the wheel and axle, lever, inclined plane, wedge, screw, toggle joint, lever, pulley, and gear wheel.

> A lever is a rigid object (usually long and narrow, like a bar or a pole) which can turn (pivot) around a fixed point (fulcrum).

> Torque is a twisting or turning to produce rotation.

(Definitions have been simplified for use in the classroom; the only precise definitions of these terms would be mathematical.)

This activity utilizes a first-class lever. A first-class lever is one in which the fulcrum is located between the resistance and effort (the input and output). A good example of a first-class lever is the teeter-totter. As you push down on one end of the lever(effort), the load (resistance) goes up on the other end. A lever remains balanced as long as there is an equal torque on either side of the fulcrum.

Although the movement of the ends of a first-class lever appear to be a simple up-and-down or back-and-forth motion, it is important to note that the ends of the lever pivot on the fulcrum, making the movement an arc, rather than a straight line. Torque is determined by the amount of force applied and the distance from the fulcrum the force is applied.

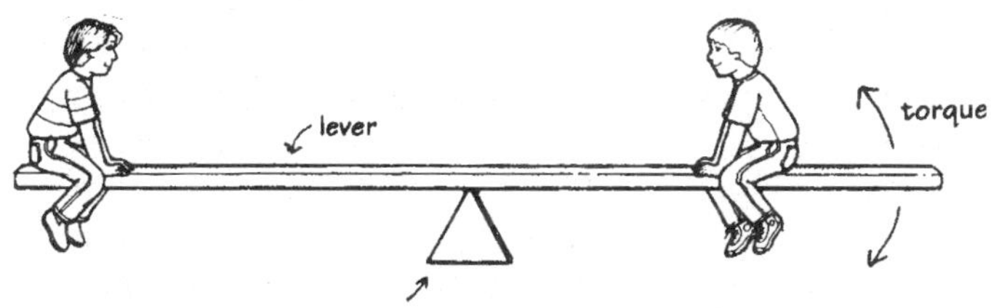

The system in this activity consists of a meter stick with a cup at one end (to which pennies will be added), using a pencil as a fulcrum. The system will be balanced when there is an equal torque on both sides of the fulcrum. As pennies are added to the cup, the torque is increased on that side of the fulcrum. In order to bring the system back into balance, the meter stick must be shifted relative to the fulcrum until balance is restored. This will result in the cup end of the meter stick being closer to the fulcrum. Note that even though the fulcrum (pencil) is not moved, the location of the fulcrum in r*elation to the meter stick is moved when the meter stick is changed to restore balance.*

Common sense might suggest that each penny added would require equal shifts of the meter stick relative to the fulcrum. This is not the case, however, because when considering leverage, it is not only important to consider the amount of force but also the position of the force relative to the fulcrum or pivot, the length of the lever arm. The greater the distance from the fulcrum, the greater the effect of the force. The resulting points do not trace out a straight line but will form a curve.

Management

1. This activity is best done in groups of four: one to add pennies, one to move the lever, one to judge if the lever is in the balanced position, and one to record the measurement.
2. Using pennies minted in or after 1982 is recommended because the composition and weight of pennies have changed several times through the years. Beginning in 1982, the cents minted have contained 2.5 percent copper and 97.5 percent zinc; they weigh 2.50 grams each. *(Information source: U.S. Mint)*
3. Make sure that students hang the cup at the 100 cm end of the meter stick. Tape, paper clips, or string may be used to attach the cup to the meter stick. They must tape the pencil to the table in a position that allows the system to function.
4. Students will need patience in balancing the meter stick and in measuring millimeters. You may wish to prepare students with practice in both of these areas.
5. Allow one class period for this activity.

Procedure

1. Discuss the *Key Question*: How must the location of the fulcrum change to keep a system in balance when weight is added to one end?
2. Distribute the materials to each group.
3. Students hang the cup from the meter stick.
4. Students tape the pencil close to the edge of a table.
5. Students balance the meter stick on the pencil with the cup hanging over the edge of the table. (The meter stick will be a pencil's thickness above the table.)
6. Students record the length of A, the distance from the fulcrum to the free end of the meter stick.
7. Add five pennies to the cup and move the meter stick until the system is balanced again.
8. Students record on the activity sheet chart the length of A and the difference between this measure and the previous length of A.
9. Repeat Steps 7 and 8 with each additional five pennies until 50 pennies are in the cup and the data have been recorded.
10. Students use data in the table to construct a broken-line graph.

Discussion

1. How did the length of A change as you added more pennies? [increased]
2. How did the difference change as the number of pennies was increased? [decreased]
3. How does the graph show changes in the length of A and the difference between each pair of succeeding data?
4. Why does it take less and less added length to balance more and more pennies?

Extensions

1. Predict and measure the length of A with 60 pennies. . . .100 pennies. [extend the graph]
2. Hang the cup from the zero end of the meter stick and repeat the activity. Construct a new chart and graph using the distance from the cup to the fulcrum instead of the length of A. How does the data change? Compare this graph with the one from the original activity.
3. Have students construct a balance using this system: Cover the numbers on the meter stick with masking tape. Balance the system and mark the location as zero. Place gram masses in the cup in increments of 5 or 10 grams. Note on the tape the location of the fulcrum and the number of grams each time.
4. Without disclosing it, give some groups pennies minted in or before 1981. (Although these pennies will vary slightly in weight among themselves, each is almost half a gram heavier than the 1982-and-later cents.) When groups obtain different data, let students work to discover the reason.

Major Conceptual Components

- Energy comes in many forms.
- Mechanical energy is the energy due to the position or the movement of something.
- A machine is a device for multiplying forces or changing the direction of forces.
- Levers help to lift loads with less effort. The force needed for lifting changes as the distance of the effort arm changes.

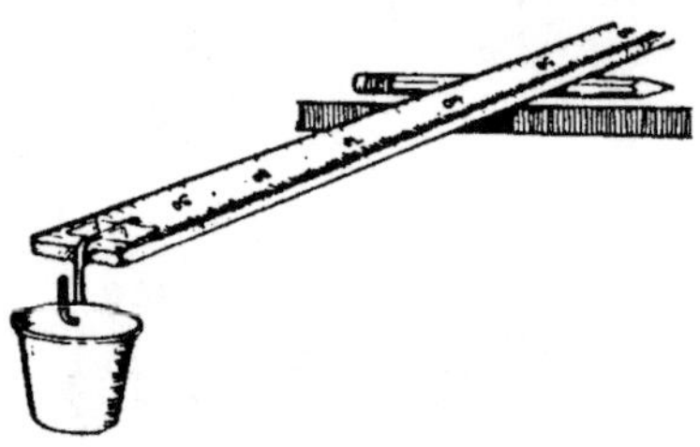

Penny Weight

Name ____________________

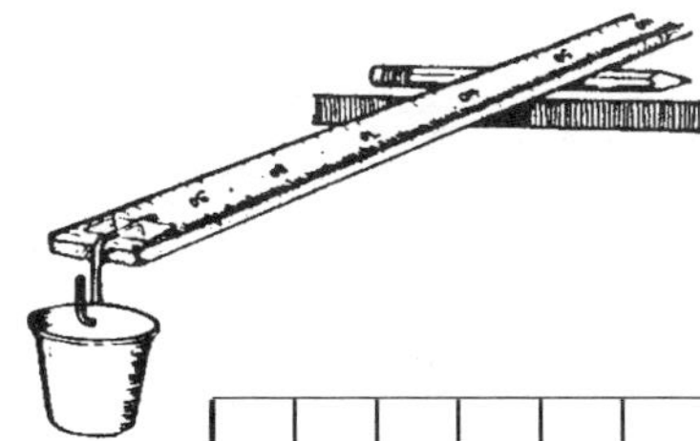

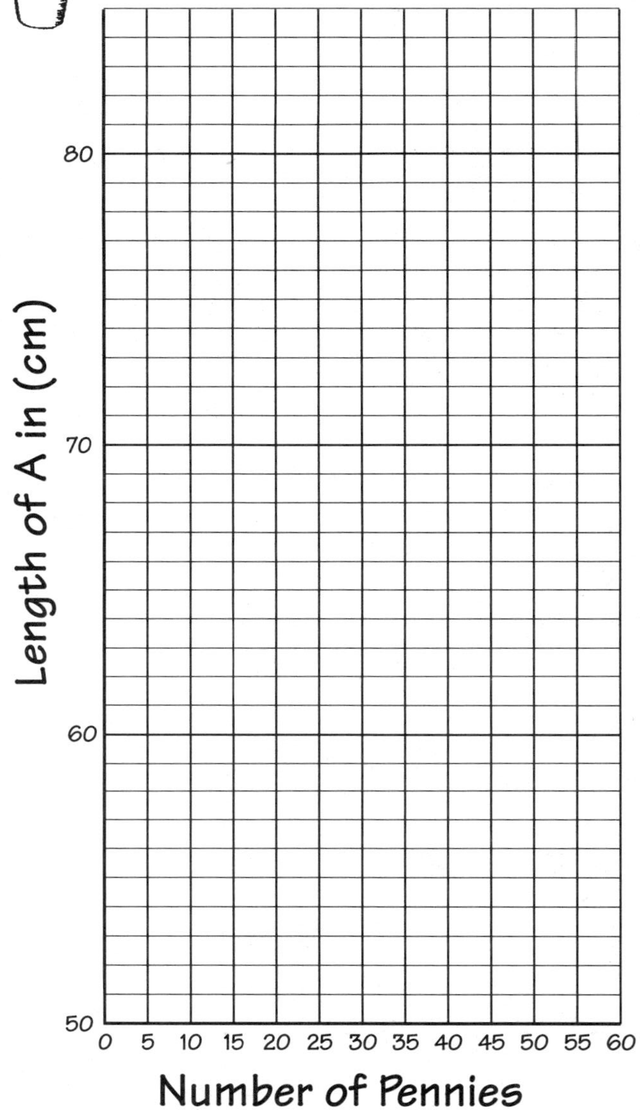

How must the location of the fulcrum change to keep a system in balance when weight is added to one end?

Number of Pennies	Length of A	Difference
0		——
5		
10		
15		
20		
25		
30		
35		
40		
45		
50		

1. Hang a cup on the 100 cm end of a meter stick.
2. Balance the meter stick on a pencil taped near the edge of a table.
3. Record on the chart the length of A, (the free end of the meter stick starting at the pencil).
4. Add five pennies and rebalance the system.
5. Record the new length of A.
6. Find and record the difference between this length of A and the previous one.
7. Complete the chart in the same way.
8. Construct the graph, using the data in the chart.

What can you infer from the graph?

____

____

____

Clever Lever 1

Topic Area
First class lever

Introductory Statement

Students will build a first-class lever and explore how it works. They will try to find a pattern relating the resistance force, resistance distance, effort force, and effort distance. They calculate the mechanical advantage of the lever and recognize its relationship to the different forces used on the lever.

Math
Using computation
Measuring
 length
 mass
Identifying number patterns
Using equations

Science
Physical science
 Simple machines

Math/Science Process Skills
Observing
Collecting and recording data
Interpreting and inferring
Applying and generalizing

Materials
Per group:
 meter stick (see *Management 6)*
 clay (small piece)
 10 pennies, minted after 1982
 masking tape
 jumbo paper clip
 hexagonal pencil

Key Questions
Part 1: On a lever, how does the size and position of a weight influence the amount of effort needed to move it?
Part 2: If you know how much a friend weighs and where he or she is sitting on a seesaw, how can you determine where to sit to balance him or her?

Background Information

A machine is a device that lets you do work with less force. It does not save you any work; it only adjusts the product of distance and force to allow you to use less effort. In scientific terms, work is defined as the product of the force on an object and the distance through which the object is moved: more simply, Work = *force x distance.* Energy cannot be created or destroyed; it can only be changed from one form to another. This is the *law of the conservation of energy*. As a consequence of this, *work in = work out.*

Simple machines conserve energy. The amount of work put in as effort is returned as work out to move the resistance. This activity focuses on one type of simple machine, the lever. All levers consist of the following components:

- A fulcrum–the pivot point for the lever arms
- A resistance (r)–the resisting force produced by the weight of the object you are trying to move
- A resistance arm–the distance from the fulcrum to the center of the mass of the resistance
- An effort (e)– the force applied to the lever to move the resistance
- An effort arm–the distance between the fulcrum and the center of the mass of the effort.

There are three classes or types of levers; *Clever Lever 1* focuses on one type, the first-class lever. The fulcrum of a first-class lever is located between the resistance (r) and the effort (e).

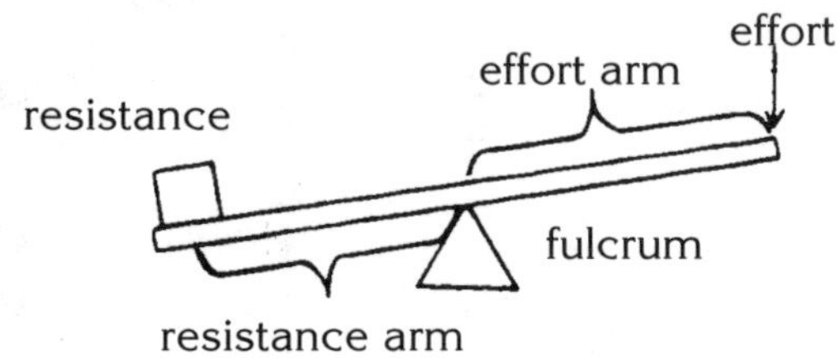

As with all simple machines, the lever trades distance for force. For example, in the following illustration, in order to raise the object with a resistance force (f_r) of 200 N, the fulcrum can be adjusted so that the effort arm is twice as long as the resistance arm. This adjustment allows an effort force (f_e) of 100 N to be used. 100 N of force with an effort arm distance of 2 units will raise 200 N of force with a resistance arm distance of 1 unit. (Notice, also, that the first-class lever can change the direction of a force.)

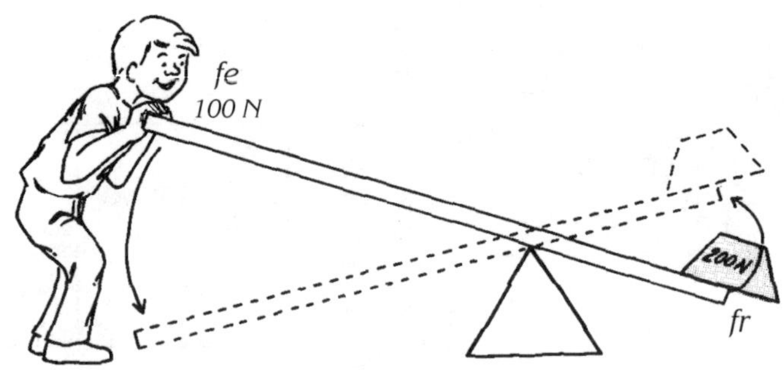

Because energy is conserved, the smaller effort force times the greater distance (d_e) to which it is applied equals the greater resistance force (f_r) times the smaller distance (d_r) the resistance moves on the other side.

$$f_e \times \mathbf{d_e} = \mathbf{f_r} \times d_r$$

If the lever reduces the amount of effort required to move the resistance, the effort will need to be applied over a greater distance. The lever makes an exchange: reduced effort for greater distance that effort is applied. The key concept here is the *law of the lever*. If the lever is balanced (equal torque on both sides), then

The effort force (e) times the distance of the effort arm	will equal	the resistance force (r) times the distance of the resistance arm

The law of the lever equation gives the measure of the amount of torque on each side of the lever. (For more information on torque, see *Penny Weight.*) The amount of torque can be calculated by multiplying the force acting on a lever arm by the distance of the lever arm. Restated, it says that when the effort arm and the resistance arm have equal amounts of torque, the lever will balance.

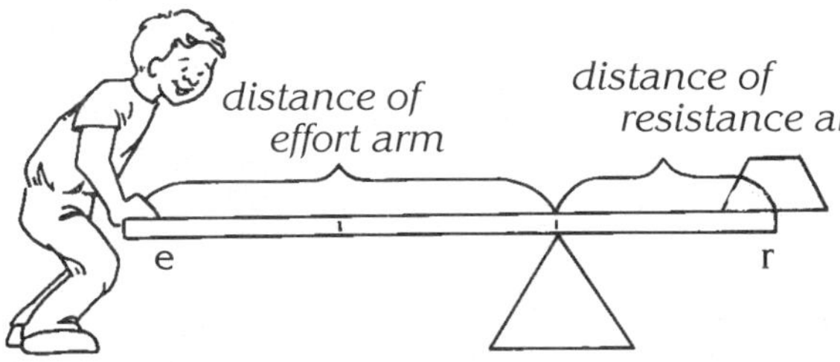

Another important concept involved in simple machines is that of mechanical advantage. The mechanical advantage (M.A.) of a machine is the number of times the machine multiplies the force you use to do a job. To find out how much the lever is helping beyond the effort it would take you to lift directly, use the formula below.

$$\text{M.A.} = \frac{r}{e} \qquad \text{M.A.} = \frac{6}{2} \qquad \text{M.A.} = 3$$

If the resistance is six newtons and you are exerting a force of two newtons, the M.A. would be three, meaning you can lift a resistance three times the effort. Another way to describe the situation is that you would use one-third the effort that you would need if you lifted the rock directly.

Management

1. Allow one class period for this activity.
2. Both parts are best done in groups of four students: one to build the balance and verify that it is level for each measurement, one to place the pennies on the left, one to place the pennies on the right, and one to record measurements.
3. The blank columns on the activity sheet, for recording the torque, should be left until after all of the other data have been recorded. (See *Procedure 6)*.
4. The composition and weight of pennies has changed several times through the years. Beginning in mid-1982, the pennies minted have contained 2.5% copper and 97.5% zinc; they weigh 2.5 grams each. *(Information source: U.S. Mint)* Have students use pennies minted after 1982 to assure a more consistent weight.
5. Have students make linear measurements to the nearest centimeter. This will allow them to discover the *law of the lever.*
6. Wooden meter sticks are recommended because the edge is thick enough to use in balancing the pennies. Since most flat plastic meter sticks are thinner than wooden ones (making it difficult to balance pennies on the edge), they can be used if you tape two of them together, being sure to use an equal length of tape at corresponding locations to maintain a balance.

Procedure

Part 1

1. Discuss the first *Key Question:* What pattern can you find between the number of pennies and the distance from the fulcrum on the resistance side and the number of pennies and the distance on the effort side? Share ideas about how it might be answered, but allow a *let's find out* atmosphere to prevail.
2. Distribute the materials.
3. Using steps 1-3 on the activity sheet, guide each group of students to assemble the lever. Help them to identify the resistance, resistance arm, effort, and effort arm and use this terminology to describe the levers they have built.
4. Follow directions on the activity sheet to fill in the resistance and effort columns. *Be sure students measure in both directions from the fulcrum, also that they measure to the middle of the pennies each time.* Distance is always measured from the fulcrum to the center of the mass of the objects used for the effort and resistance forces.
5. Discuss the concept of mechanical advantage (M.A.). (See *Background Information.*). Students compute the M.A. and enter data in the appropriate column.
6. Encourage students to discover the law of the lever. Guide them to see patterns in the data between the number of pennies and the distance from the fulcrum. Allow students some time to explore the possibilities. Ask the first *Key Question* again. If students become too frustrated, you might suggest looking at one row at a time and trying all four basic computation operations. When students realize that the force and distance should be multiplied, encourage them to fill in the blank columns to record their data. Introduce the term torque and use it to label the columns *Torque of r* and *Torque of e.* The fact that each pair of numbers in these two columns is equal (approximately) is the law of the lever.

Part 2

1. Discuss how the second *Key Question* can be answered using the law of the lever. Have students use any reasonable figures in the law of the lever and find the amount of torque, as in the chart.

Discussion

1. With a balanced lever, what pattern can you find between the number of pennies and the distance from the fulcrum on the resistance side, and the number of pennies and distance on the effort side? [$f_r \times d_r = f_e \times d_e$]
2. Was the product of force and distance exactly equal on both sides of every lever? If not, what are some likely causes for the differences? [Most measures will be close, discrepancies may be cause by inaccurate measures or not leveling the balance.]

3. Discuss the solution to the *Key Question.* [Use the law of the lever]
4. What pattern do you see between the mechanical advantage and the number of pennies used in each arrangement? [ratio of resistance pennies to effort pennies]
5. What does the mechanical advantage tell you about the force made by each penny on the effort side compared to each penny on the resistance side? [the fraction of force generated by each penny}

Extensions

1. Make a mobile using four pieces of dowel (pencils or straws may be substituted) of various lengths and five objects of various weights. How can you use what you learned to make the dowels hang horizontally?
2. Make a collage of pictures or cartoons showing some first-class levers. [seesaw, balance]

Major Conceptual Components

- Energy cannot be created or destroyed.
- The amount of energy or work put into a machine will come out of the machine in some form.
- Energy is used to accomplish work.
- Work is done when something is moved.
- The amount of work done is determined by how much force was used to move the object and how far the object was moved.
- Mechanical advantage is a comparison of force produced by a machine compared to the force applied to the machine.
- Levers help to lift loads with less effort. The force needed for lifting changes as the distance of the effort arm changes.

Name ____________________

Clever Lever 1

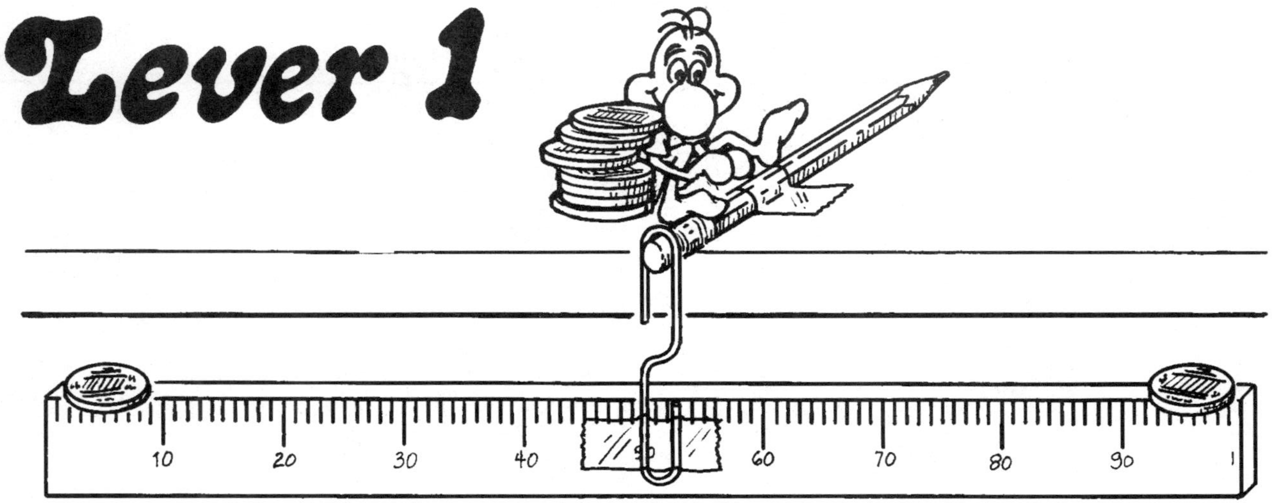

1. Bend the paper clip into an "S" and tape its SMALLER hook to the meter stick at the 50 cm mark.
2. Tape a pencil on top of the table so that the end with the eraser sticks out about an inch beyond the edge.
3. Hang the meter stick on the pencil by the big hook of the paper clip. The meter stick should hang level. If it is not level, place a small piece of clay on the bottom edge of one end until it is. Be patient!
4. Balance one penny on the resistance (left) side of the lever, with its center 48 centimeters from the fulcrum (pencil). Then, balance the lever with 1 penny on the effort (right) side. How many centimeters is the center of this penny from the fulcrum? Record the length of this effort arm on the chart.
5. Complete the columns for resistance and effort in the same way.
6. Compute the mechanical advantage (M.A.) for each line.

Name ______________________

Clever Lever 1

What pattern can you find between the number of pennies and the distance from the fulcrum on the resistance side, and the number of pennies and distance on the effort side?

Left Side - Resistance			Right Side - Effort			
Resistance force - pennies r	Resistance arm length (cm)			Effort arm length (cm)	Effort force - pennies e	Mechanical Advantage r/e
1	48				1	
1	48				2	
1	48				3	
1	48				4	
1	48				6	
1	48				8	
2	24				1	
2	24				3	
2	24				4	
2	24				6	
3	12				1	
3	12				2	
3	12				4	
3	12				6	
4	6				1	
4	6				2	
4	6				3	
4	6				6	

If you know how much a friend weighs and where he or she is sitting on a seesaw, how can you determine where you need to sit to balance it?

Sandbagging the Seesaw

Topic Area
First-class lever

Introductory Statement
Students will explore the properties of effort, resistance, and torque using a seesaw as a first-class lever.

Math
Using computation
Measuring
- length
- weight

Using calculators
Applying a formula

Science
Physical science
- Simple machines

Math/Science Process Skills
Observing
Predicting
Collecting and recording data
Making and testing hypotheses
Applying and generalizing

Materials
Per group:
- 1 - 2" x 6" x 8' board
- 1 - 1" x 1" x 6" block of wood (scrap lumber)
- 2 - 1.5 meter tape measures
- 1 brick
- 3 nails
- hammer
- tape
- bathroom scale
- burlap sack (See *Management 7)*
- sand
- markers
- string

Key Questions
Part 1:
If one sandbag character is placed on the seesaw, how can you know beforehand where to place another sandbag so the seesaw will balance?
Part 2:
How can you determine the weight of a sandbag using the seesaw?

Background Information

A seesaw is an example of a first-class lever in which the fulcrum is located between the resistance and the effort. In this activity, the resistance force (f_r) will be the sandbag which is first placed on the seesaw (the lever). The effort force (f_e) will be the sandbag which is later placed on the seesaw to achieve a balance the seesaw. This sandbag's position can be manipulated while the resistance sandbag must remain stationary.

The *law of the lever* is fundamental to the understanding of this activity: if the lever is balanced, then

The effort force (f_e) times the distance of the effort arm	will equal	The resistance force (f_r) times the distance of the resistance arm

or, stated mathematically,

$$f_e \text{ x distance of effort arm } = f_r \text{ x distance of resistance arm}$$

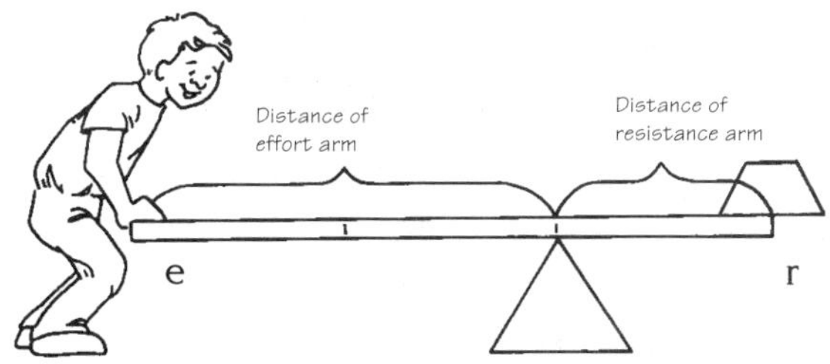

Keep in mind that the distance (length) of the effort arm extends from the fulcrum to the center of gravity of the effort force; the same is true of the resistance arm and the resistance.

In *Part 2* of this activity, students will use symbols such as the following:

f_e = force of the effort
f_r = force of the resistance
d_e = distance of the effort arm
d_r = distance of the resistance arm
T_e = torque of the effort
T_r = torque of the resistance

The use of letters with subscripts is a common practice in advanced mathematics and science. The symbols are simply abbreviations which allow us to write equations related to a concept. In this particular case, they allow us to show the equality that exists in the law of the lever.

When determining the effort and resistance forces, weight instead of mass is used. There is a difference between weight and mass. Weight refers to the force due to gravitational attraction upon a body. Mass is the quantity of matter a body contains without regard to where it is located (altitude, atmospheric pressure, relation to earth's magnetic field, etc.) For example, the mass of an object would be the same on earth and in outer space; its weight would be considerably different.

The metric measurement of weight is in newtons (N), and metric measurement is more common in the scientific realm; as a result, most AIMS activities follow the usage of the metric system to help students become comfortable with scientific terms and procedures.

It is scientifically correct to measure torque in newton-meters (Nm); however, since bathroom-type scales calibrated in newtons may not be available, the following conversions may be used to determine the number of Newtons using scales which are calibrated in pounds or kilograms.

9.8 x kilograms = number of newtons
4.45 x pounds = number of newtons

Instead of making the above conversions in this activity to express the torque in newton-meters, you may prefer to retain the unit of measure on the scale you use; if so, the torque will be expressed in pound-meters (lbm). Although such results will not be scientifically appropriate, the proportions will be correct because you are using the law of the lever with the same units.

$$\boxed{f_e \times d_e = T_e} \quad = \quad \boxed{f_r \times d_r = T_r}$$

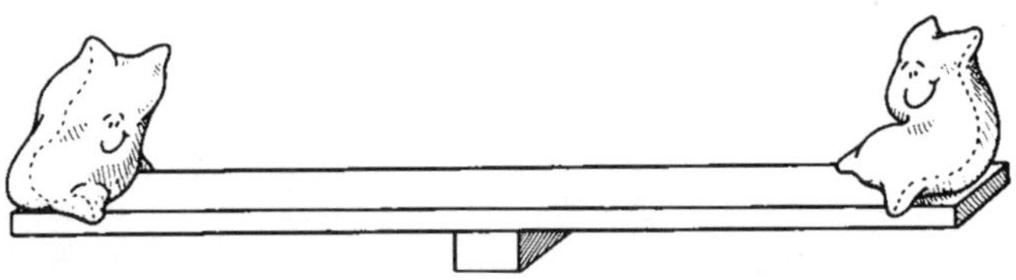

An application of the law of the lever to the action of a seesaw is this: the resistance (the resisting force) is the product of the weight of the sandbag being lifted (in newtons) and the distance (length) of the resistance arm (in meters). The result is sometimes called the *resistance torque* or *torque of the resistance* (T_r). The effort (or effort force), on the other hand, is the product of the amount of force used to push down on the seesaw and the effort arm. The result is sometimes called the *effort torque* or *torque of the effort* (T_e). When the seesaw is level or balanced, the resistance torque and the effort torque are equal.

For more discussion on levers and torque, refer to *Background Information* in *Penny Weight* and *Clever Lever 1*.

Management
1. This activity should follow *Clever Lever 1* in which students become familiar with first-class levers.
2. *Sandbagging the Seesaw* has two parts, *Part 1* is a directed application of the law of the lever as students find the distance of the effort arm. In *Part 2*, students use the law of the lever to find out the weight of a sandbag. *Part 2*, in which students find the weight of a sandbag, has two difficulty levels: *Level 1* leads students through a variation of *Part 1* with the same chart, while *Level 2* provides more advanced students with a less-structured activity. *Part 2, Level 2* may be used as an assessment tool for the activities on first-class levers.
3. If a seesaw is available on the school grounds, it may be used instead of building the one described. Make certain the seesaw is balanced before beginning the activity.
4. If no seesaw is available, build one by placing the brick on the floor with its widest side forming the base of the seesaw's fulcrum. Place the small piece of wood on top of the brick. Balance the board on top of the small piece of wood. Nail the two pieces of wood (the board and the small piece) together. When the lumber is balanced on top of the brick, the seesaw is complete.

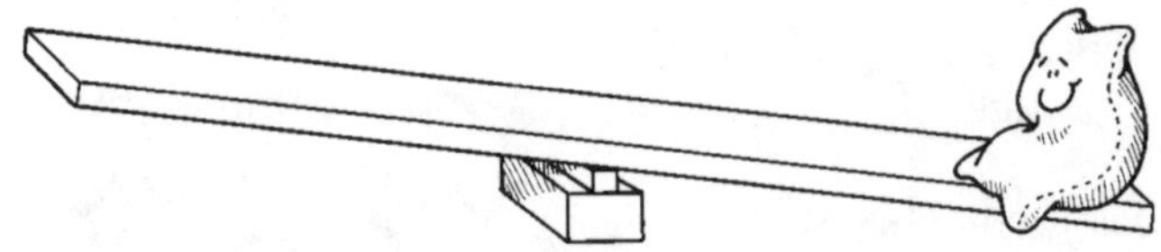

5. Tape the measuring tapes on top of the seesaw, with zero at the fulcrum and extending in both directions to the ends of the board.
6. Sandbags will be placed on the seesaw so that the mark on the meter tape is directly under the bag's center of gravity.
7. If burlap bags are not readily available, heavy-duty garbage bags may be used. Several bags may be required, one inside the other, to withstand the weight of the sand. Other things may also be substituted for the sand: dirt, rocks, books, etc.
8. It may be necessary to review how to convert centimeters to meters before beginning the activity. Example:

$$127\,\cancel{cm} \times \frac{1\ m}{100\ \cancel{cm}} = \frac{127\ m}{100} = 1.27m$$

Procedure
Part 1
1. Prior to doing the actual activity, divide students into groups. Distribute the burlap bags and direct them to design a sandbag character. If desired, the students may decorate their sandbags with items brought from home. After the sandbags have been decorated, fill them with sand and tie them securely.
2. To begin the activity, have students weigh their sandbags. Instruct them not to tell the weights because determining the sandbags' weights is the goal of *Part 2* of this activity.
3. Select a sandbag character to represent the resistance force. This will be Bag A. The other sandbags will represent the effort forces. Tell the students the weight of Bag A and place its center of gravity 60 cm (0.6 m) from the fulcrum of the seesaw. Students should record this data, noting the units used, on their activity sheet. Direct them to compute the resistance torque and record it. Have the students who made Bag A disperse to other groups.
4. Since the effort torque and the resistance torque should theoretically be equal in order to balance the seesaw, students can now record the amount of effort torque.

5. Using the various weights of the other sandbag characters, students should be able to discover where to place their characters in order to balance the seesaw. To compute the distance of the effort arm (fulcrum to the center of gravity for the effort force), divide the effort torque by the effort force (weight). Record this value, including the units used. This is the predicted distance from the fulcrum.
6. Have the groups check their predictions by placing their sandbag characters at the predicted location. The students should then move the characters, if necessary, to make the seesaw balance.
7. Have each group measure and record the actual distances in the boxes on their record sheets.
8. Direct students to calculate the percent of error.

Part 2

1. Discuss the question at the top of the activity sheet. Give the students time to think of ways they can find the answer. Ask students how they can use the seesaw as a scale.
2. Tell the students that a sandbag character–Bag B (a different one than used for the resistance force in *Part 1)* will be placed at a given distance from the fulcrum. Do not disclose the weight of this sandbag. This sandbag should be placed close enough to the fulcrum so that the lightest sandbag will be able to balance. Group members who made this sandbag should be dispersed into other groups so they can do the problem-solving component of this activity.
3. Bag B is placed on the seesaw.
4. *Level 1*: One by one, each group places their sandbag on the effort arm and balances Bag B that is already in place. The groups note and record (under *Your Bag)* the distance from the fulcrum to the center of gravity for their bags. Using the law of the lever, students should compute and record the effort torque, resistance torque, and predicted weight of Bag B.
   Level 2: Direct students to predict Bag B's weight using the law of the lever.
5. *Both levels:* Measure and record the actual weight of Bag B.
6. Determine the percent of error.

Discussion

1. If you weren't correct the first time you tried to balance Bag A, how did you make adjustments? How did you know to do this?
2. What are some reasons your predictions were close but not "exact"?
3. How does this relate to the pattern you saw in *Clever Lever 1*?
4. Why is the seesaw a first-class lever?
5. Identify the effort arm and the resistance arm on the seesaw.

Extensions

1. Find two students who don't mind disclosing their weights. Have one stand at a certain location on the seesaw. Ask the class to determine where the other student should stand in order to balance the seesaw.
2. Have students study a triple-beam balance. By observation, have them determine and explain how it works.
3. Design a system in which you find the weight of a heavy object by using the law of the lever.

Major Conceptual Components

- Energy is used to accomplish work.
- Work is done when something is moved.
- The amount of work done is determined by how much force was used to move the object and how far the object was moved.
- Levers help to lift loads with less effort. The force needed for lifting changes as the distance of the effort arm changes.

SANDBAGGING THE SEESAW

Name ____________________________

Part 1

If a sandbag character (Bag A) is placed on a seesaw, how can you know beforehand where to place your sandbag so the seesaw will be balanced?

1. Record your bag's weight on the picture below, but do not tell anyone what it is.
2. Record Bag A's weight and the distance of the resistance arm (60 cm).

Your Bag's weight	Bag A's weight

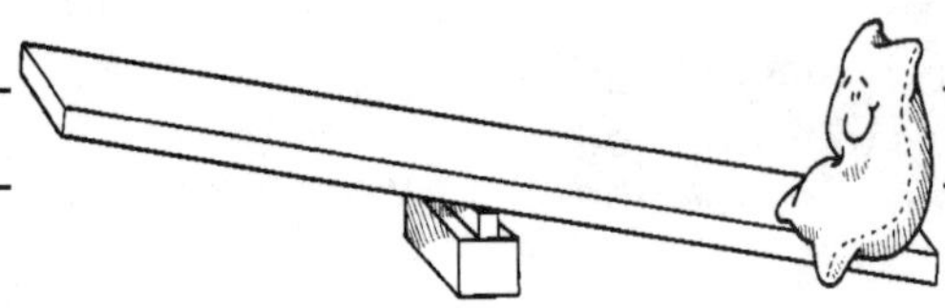

Predicted distance of effort arm	Distance of resistance arm

3. Fill in the measures you know on the chart below. Complete the calculations to predict the distance of the effort arm. This will tell you where to place your sandbag to balance the seesaw.

Your Bag				Bag A		
Effort Force (Your Bag's Weight) in ______	Predicted distance of effort arm in meters	Effort Torque in ______		Resistance Torque in ______	Distance of resistance arm in meters	Resistance Force (Weight of A) in ______
f_e ×	d_e =	T_e	=	T_r =	d_r ×	f_r

distance of Effort Arm

4. Go to the seesaw and see how close your prediction is. Record the distance of the effort arm when you balance Bag A.

5. Determine the percent of error. Find the difference between your prediction and the actual distance of the effort arm. Complete the ratio below.

$$\frac{\text{difference between predicted and actual distance}}{\text{actual distance of effort arm}} = \frac{\boxed{}}{\boxed{}} \times 100 = \boxed{}\ \%$$

SANDBAGGING THE SEESAW

Name ______________________________

Part 2, Level 1

How can you determine Bag B's weight using a seesaw?

1. Use the seesaw to get the measurements you need. Use your measurements to estimate Bag B's weight.

Your Bag			Bag B		
Effort Force (Your Bag's Weight) in ______	Distance of Effort Arm in meters	Effort Torque in ______	Resistance Torque in ______	Distance of Resistance Arm in meters	Predicted Resistance Force (Weight of B) in ______
f_e ×	d_e =	T_e =	T_r =	d_r ×	f_r

Actual Weight

2. Note here Bag B's actual weight.

3. Now find the percent of error, using Bag B's actual weight and the difference between your estimate and the actual weight.

$$\frac{\text{difference between predicted and actual weight}}{\text{actual weight of Bag B}} = \frac{\boxed{}}{\boxed{}} \times 100 = \boxed{}\ \%$$

Does it matter which side the effort or resistance is on? Explain.

____

____

____

How could you get a more accurate measure of weights using the seesaw? Explain.

____

____

____

SANDBAGGING THE SEESAW

Name ______________________

Part 2, Level 2

How can you determine Bag B's weight using a seesaw?

1. Use the seesaw to get the measurements you need. Use the law of the lever to estimate Bag B's weight.

Your Bag's Weight

Predicted Weight of Bag B

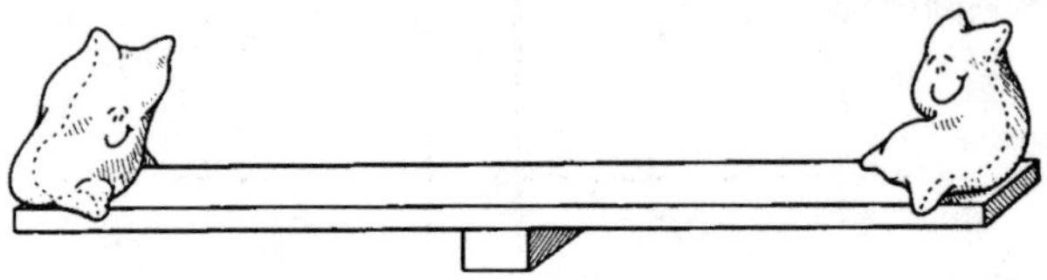

Distance of effort arm

Distance of resistance arm

Do your work here.

2. Determine the percent of error. Find the difference between your prediction and Bag B's actual weight. Complete the ratio below.

$$\frac{\text{difference between predicted and actual weight of Bag B}}{\text{actual weight of Bag B}} = \frac{\square}{\square} \times 100 = \square\ \%$$

Does it matter which side the effort or resistance is on? Explain.

____

____

____

How could you get a more accurate measure of weights using the seesaw? Explain.

____

____

____

Clever Lever 2

Topic Area
Second-class levers

Introductory Statement
Students will build a second-class lever and explore how it works. Relationships between resistance force, resistance arm distance, effort force, and effort arm distance will be determined and result in the development of a formula. Students will calculate the mechanical advantage of the lever and recognize its relationship to the different forces used on the lever.

Math
Measuring
  length
  mass
Using computation
Identifying number patterns
Developing and using formulae
Graphing

Science
Physical science
  Simple machines

Math/Science Process Skills
Observing
Collecting and recording data
Interpreting data
Drawing conclusions
Applying and generalizing

Materials
For each group:
  1 wooden or triangular meter stick
  3 jumbo paper clips
  paper cup
  masking tape
  pencil
  2 metric rulers
  1 thin rubber band
  masses or other small dense objects (clay, pennies, etc.)

Key Question
To use a wheelbarrow with the least amount of effort, where do you put the load (resistance) and where do you grab the handles (effort)?

Background Information
The lever, as with all simple machines, conserves energy. The Law of Conservation of Energy states energy cannot be created or destroyed; it changes form but does not cease to exist. This means that the amount of energy put into the machine equals the amount of energy put out by the machine (ignoring the amount lost to friction). Energy is measured as work.

Work In = Work Out

Since work is the product of force multiplied by distance, a little force put in over a long distance on the effort arm of a lever will move a great force for a short distance on the resistance arm of a lever.

A second-class lever has the resistance (load) between the fulcrum and the effort. In a second-class lever, the resistance arm and the effort arm are on the same side of the fulcrum. Because the effort arm of a second-class lever is always longer than the resistance arm, the mechanical advantage will always be greater than 1. This means that you use less force to move the load than if you lifted it directly. The distance the effort arm moves will be greater than the distance the resistance arm moves. Examples of second-class levers include bottle openers, nutcrackers, wheelbarrows.

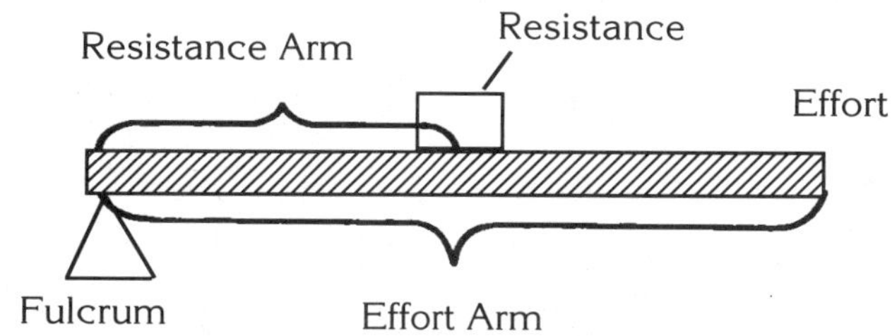

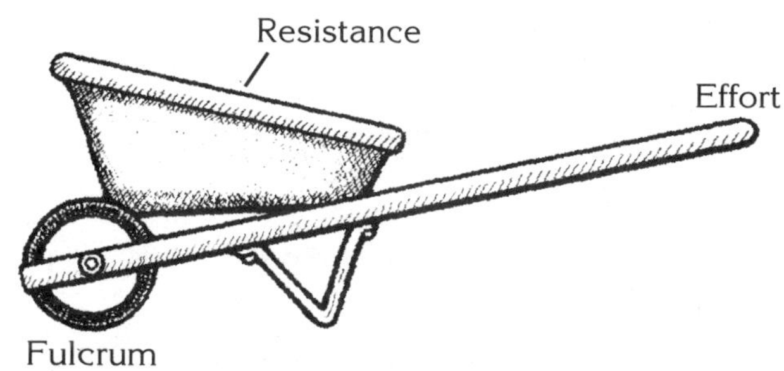

For more information on levers, refer to the *Background Information* in *Clever Lever 1*.

Management
1. Allow two class periods for the activity with discussion and graphing. Have the students build the lever and take their measurements on the first day. The second day can then be used to complete the graphs and discuss conclusions.
2. This activity is best done in groups of four: one to construct the lever, one to move the resistance, one to measure, and one to record.
3. Before the activity, the teacher must determine the appropriate amount of mass to put in the cup. To do this, stretch a rubber band of the type to be used by the class to its maximum length and measure. Then open a paper clip to an S shape and hang a paper cup from the clip by poking a hole near the rim of the cup. Hang the clip and cup from the rubber band using the other curve of the S. Add masses or other dense objects to the cup until the rubber band is stretched to about 2/3 its maximum length. The teacher may choose to put the appropriate amount in each cup before the

activity or allow the students to do it at the beginning of the activity.

4. In this activity, the measurement of the force of the effort is the measure of the stretch of the rubber band. This stretch is to be measured in millimeters. Because students may become confused to have both distance and force measures using the same units and because a millimeter is not a force unit, non-standard units called rubber band units (rbu) will be used for the force measurement. Although students will actually measure the stretch in millimeters, rbu's will be recorded to differentiate between the force and distance measures.

Procedure

1. Discuss the *Key Question.*
2. Distribute the materials to each group.
3. Have the students build the lever and complete the activity by following the instructions on the student pages.
4. Using the data, have the students make the graphs.
5. Direct the students to use their graphs to draw conclusions and make predictions which may be tested with their levers.

Discussion

Part 1 - Force

1. What happens to the length of the rubber band as the resistance gets farther from the fulcrum? [It gets longer] What does this tell us about the amount of force? [More effort force is required]
2. Describe the lines on the force graph? [They both go up and they are always the same distance apart]
3. What different information does each line give you? [total stretch, net stretch]
4. Why are the two lines always the same distance apart? [The distance apart is the length of the rubber band when there is no resistance on the lever]
5. Use the graph to predict how long your rubber band would be if the resistance were 500 mm from the fulcrum and test your prediction. [Answers will vary. Have students discuss the method of prediction and their accuracy. Discuss reasons for discrepancies in predictions and actual measures]
6. What does the graph show about the distance the resistance is from the fulcrum and the effort required to hold up the resistance? [As the resistance is farther from the fulcrum it requires more effort to hold the resistance]
7. If you want to lift a very heavy object with very little force, where would you put the resistance and the effort on the lever? [resistance close to the fulcrum, effort far away from the fulcrum] Explain your answer.

Part 2 - Distance

8. What happens to the distance the effort moves as the resistance gets farther from the fulcrum? [The distance the effort moves gets less]
9. Describe the line on the force graph? [It goes down] What information does this tell you?
10. Use the graph to predict how long your rubber band would be if the resistance were pulled down 20 mm at the position 500 mm from the fulcrum. Test your prediction. [Answers will vary. Have students discuss the method of prediction, and their accuracy. Discuss reasons for discrepancies in predictions and actual measures]
11. What does the graph show about the distance the resistance is from the fulcrum and the distance the effort moves? [As the resistance moves farther from the fulcrum, the effort moves less]
12. A load in a wheelbarrow needs to be lifted only a little bit to allow you move the wheelbarrow. How far do you need to lift the handles to lift the load this little bit? [A lot farther than the load]
13. Compare the two graphs and discuss their differences. [The line on the force graph goes up, while the line on the distance graph goes down]
14. What conclusions can you make about what happens to the effort as the resistance gets farther from the fulcrum? [The effort required to hold the same resistance gets greater, but the distance the effort moves for the resistance movement becomes less]
15. Discuss answers to *Key Question.* [Place the load very close to the wheel, the fulcrum, and grab the handles as near the ends as possible]
16. The lever you built is called a second-class lever. How is this lever different from the lever you built in *Clever Lever 1* [The resistance is between the fulcrum and the effort]
17. What other second-class levers can you name? [Nutcracker, bottle opener, wheelbarrow]

Extensions

1. Bring a wheelbarrow into the class and have students test their conclusions. Allow them to experience the amount of effort required to move a student (the resistance) who is sitting in various locations in the wheelbarrow. Students will feel the differing amounts of effort, and can see how much distance it takes to move a student that is close to the wheel. Have students identify the effort, effort arm, resistance, resistance arm, and fulcrum.
2. Have students find other examples of second-class levers at home and bring them in to explore.
3. Have the students revisit the activity moving the resistance 10 mm (1 cm) at a time and see how the new graph differs from the original.
4. Do the activity with spring scales and calculate actual mechanical advantage.

Major Conceptual Components

- Energy cannot be created or destroyed.
- The amount of energy or work put into a machine will come out of the machine in some form.
- Energy is used to accomplish work.
- Work is done when something is moved.
- The amount of work done is determined by how much force was used to move the object and how far the object was moved.
- Mechanical advantage is a comparison of force produced by a machine compared to the force applied to the machine.
- Levers help to lift loads with less effort. The force needed for lifting changes as the distance of the effort arm changes.

Clever Lever 2

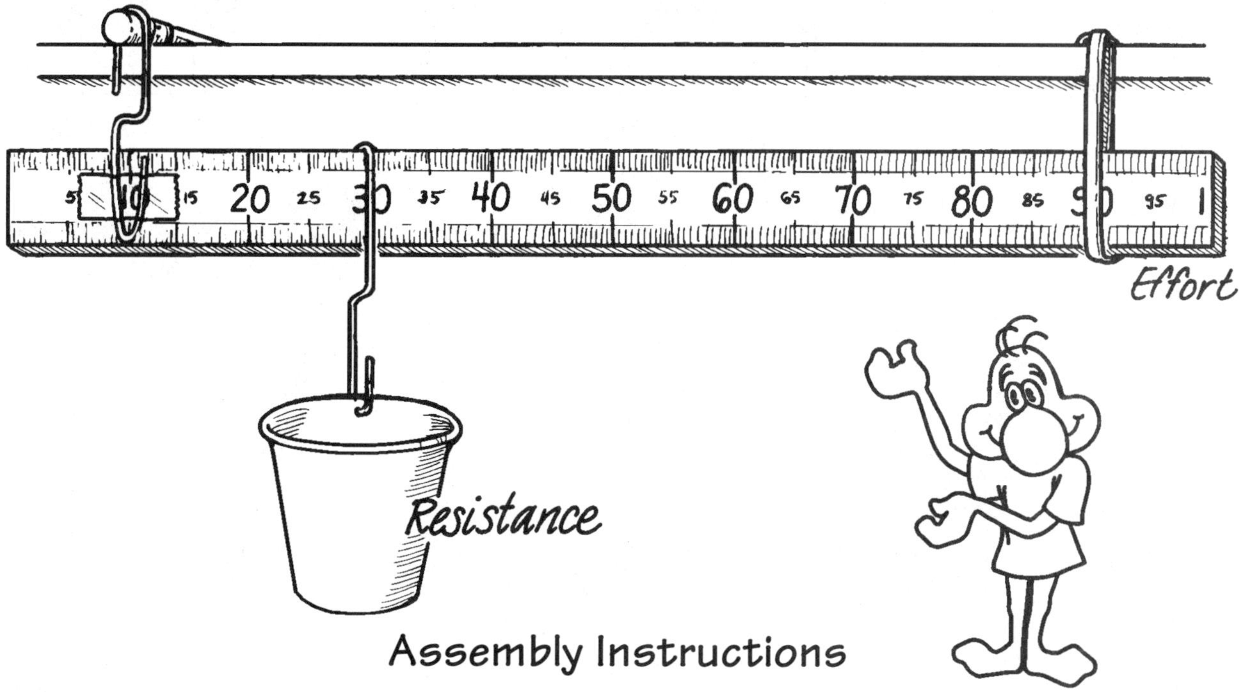

Assembly Instructions

1. Open a paper clip into an S, and tape it perpendicular to a meter stick at the 10 cm mark. Tape it so that the larger end of the S stick out above the top of the meter stick.

2. Tape a pencil to the top of a table so that the end with the eraser sticks out about 2 cm.

3. Hang the meter stick by the end of the paper clip on the end of the pencil. This is the fulcrum of the lever.

4. Slip the rubber band into one of the hooks of another paper clip. Slide the rubber band onto the end of the meter stick opposite the pencil until it is at the 90 cm mark. Tape the paper clip to the top of a the table so it supports the meter stick by the rubber band. Make sure the rubber band is still at the 90 cm mark before you tape it down. This is where the effort on the lever will be measured.

5. Bend a third paper clip into an "S". Poke the smaller hook into a paper cup just below the rim. Add weight to the cup as instructed by the teacher. By hanging the big hook on the meter stick you will be putting the resistance on the lever.

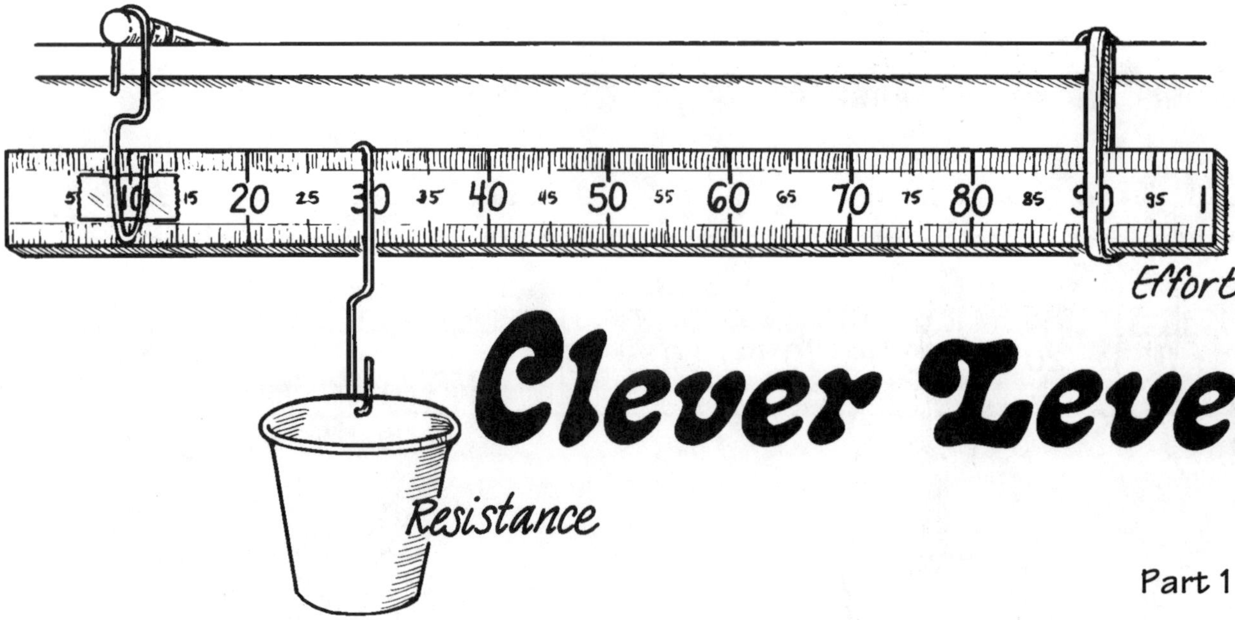

Clever Lever 2

Part 1 - Force

1. Measure the length of the rubber band before adding the resistance. Measure to the nearest millimeter and record the length on the chart beside total stretch and under Lever with no Resistance.

2. Hang the cup(resistance) with its mass at the fulcrum. Measure and record the total stretch of the rubber band and calculate the net stretch (total stretch minus stretch with no resistance). Although you will read these measurements in millimeters, they will be recorded as rubber band units (rbu) to distinguish them from distance measures.

3. Move the resistance 200 mm, measure and record the total stretch, and calculate the net stretch.

	Lever with no Resistance	Resistance's Position on the Meter Stick				
		10 cm	30 cm	50 cm	70 cm	100 cm
		Resistance's Distance from the Fulcrum				
		0 mm	200 mm	400 mm	600 mm	800 mm
Total Stretch	rbu	rbu	rbu	rbu	rbu	rbu
Net Stretch	0 rbu	rbu	rbu	rbu	rbu	rbu

Graphing

1. Determine a scale for the effort variable on the graph. Be sure to consider your longest stretch measurement.

2. Make a line graph using the values for the total stretch from your data.

3. Graph the net stretch as a second line on the graph. Label both lines.

Clever Lever 2

Force Graph

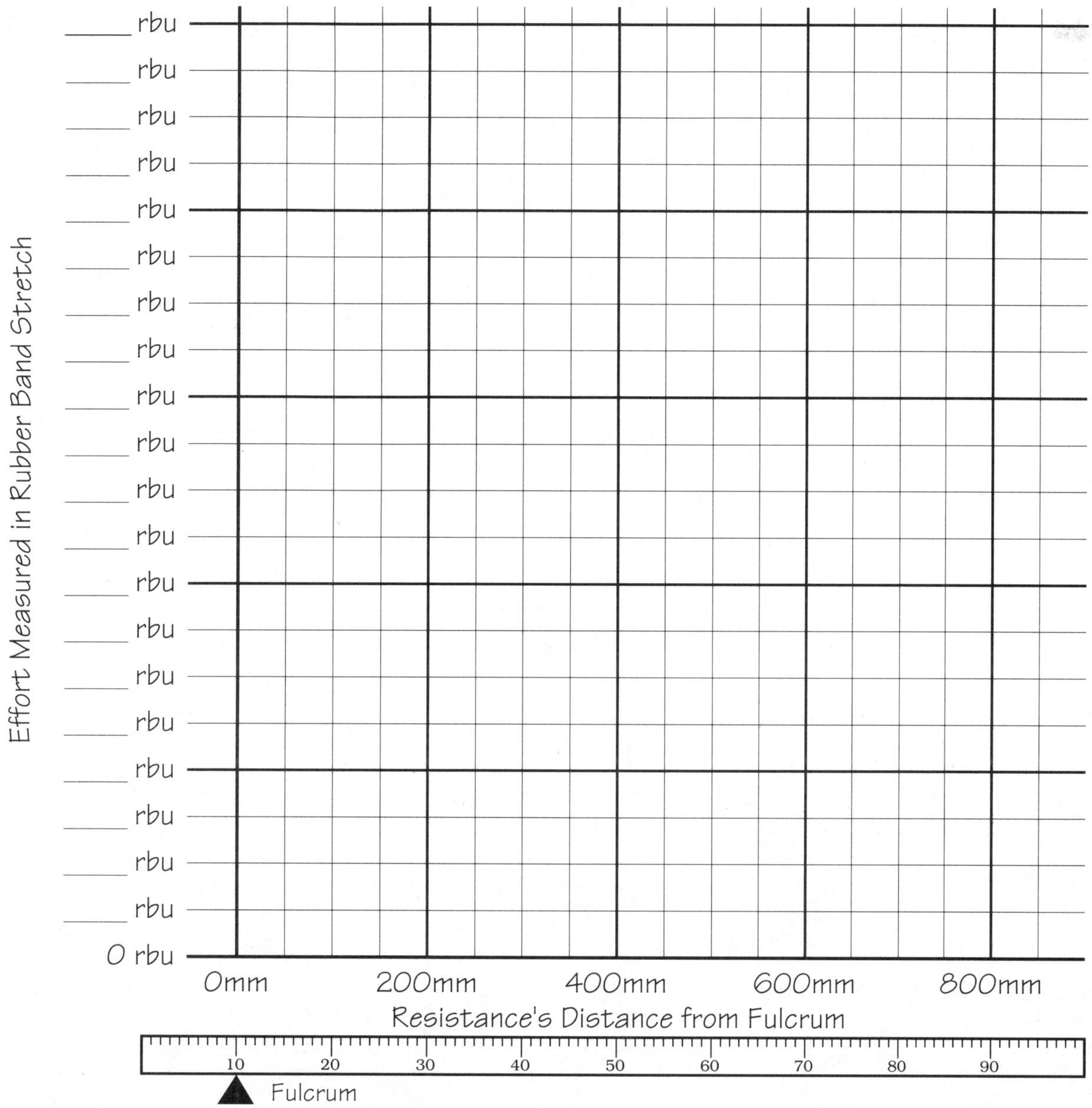

What conclusions can you draw from the graph?

Clever Lever 2

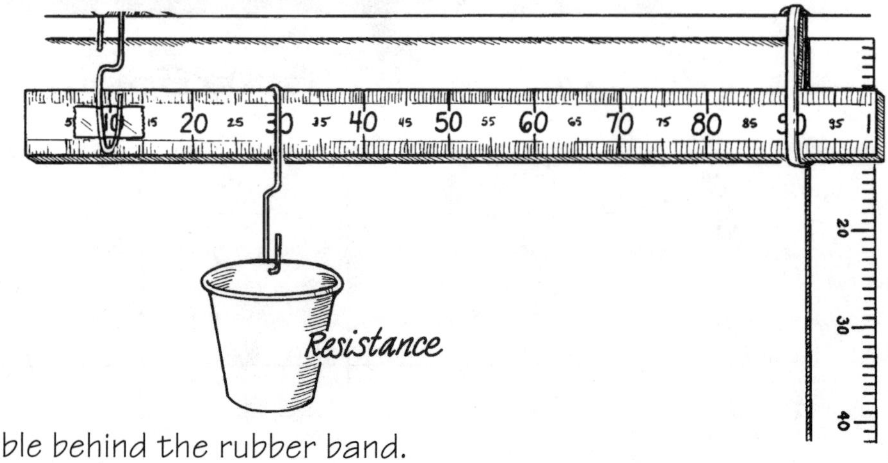

Part 2 - Distance

1. Tape a ruler vertically to the table behind the rubber band.

2. Measure and record as the zero position the distance from the top of the ruler to the top of the meter stick.

3. Empty the cup and hang it at the 30cm mark, 200mm from the fulcrum.

4. Hold the other ruler vertically behind the meterstick at the first position, 200mm from the fulcrum.

5. Pull the cup down until the top of the meter stick moves down 20mm at this position.

6. Hold the stick in this position and measure the distance from the top of the meter stick to the top of the ruler at the rubber band's position.

7. Record the distance the effort has moved from the zero position. Be sure to subtract the starting measurement found in step 2.

8. Complete the chart by repeating steps 4 to 7, changing the resistance's distance from the fulcrum each time.

	Resistance's Position on the Meter Stick			
	30 cm	50 cm	70 cm	100 cm
	Resistance's Distance from the Fulcrum			
	200 mm	400 mm	600 mm	800 mm
Distance Effort Moves	mm	mm	mm	mm

Graphing

Make a line graph using the values for the distance the effort moves from your data.

Clever Lever 2

Distance Graph

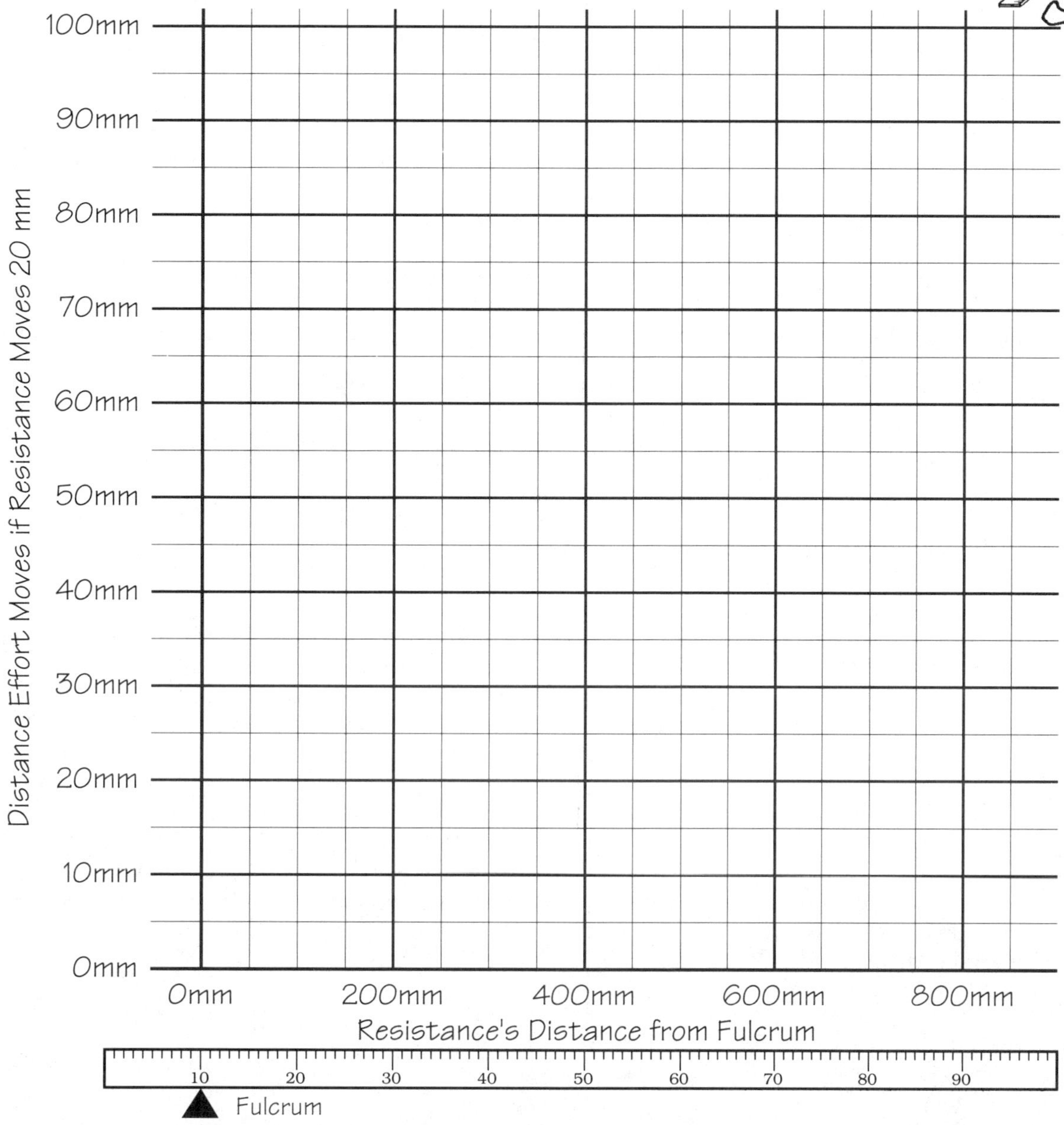

What conclusions can you draw from the graph?

What conclusions can you you draw by comparing the distance and force graphs?

Clever Lever 3

Topic Area
Third-class levers

Introductory Statement
Students will build third-class levers and explore their properties of resistance, effort, ideal mechanical advantage and actual mechanical advantage.

Math
Measuring
  length
Graphing
Using calculators

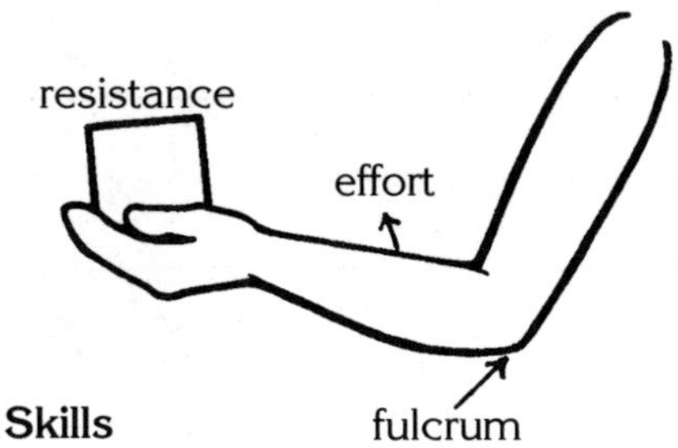

Science
Physical science
  Simple machines

Math/Science Process Skills
Observing
Collecting and recording data
Interpreting data
Comparing
Applying and generalizing

Materials
Per group:
  wooden or triangular meter stick
  3 large paper clips
  2 metric rules
  masking tape
  pencil
  2 thin rubber bands
  scissors

Key Question
How does a lever work differently when your push or pull (effort) is closer to the fulcrum than the thing you are trying to move (resistance)?

Background Information
In a third-class lever, the effort and resistance are on the same side of the fulcrum with the effort between the fulcrum and the resistance.

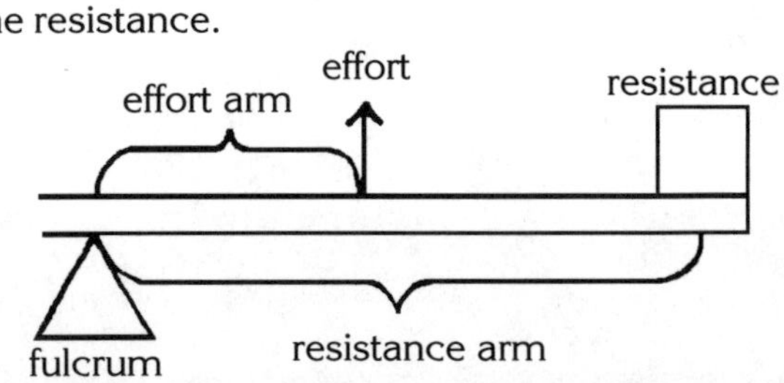

The third-class lever only allows the effort to lift a resistance of less force. However, the resistance will be moved a greater distance than the effort. Some examples of third-class levers are fly swatters, brooms, paintbrushes, golf clubs, and fishing rods.

A third-class lever allows someone with a relatively large amount of force to move something a great distance without moving very far themselves. A fly rod is a good example: The wrist of the fisherman is the fulcrum, the hand is the effort and the tip of the rod is the resistance. While the fisherman may move the rod with a great deal of wrist strength, the tip of the rod can be easily held still with a finger. As a result, although the fisherman's hand moves little distance, the tip of the rod moves through a great arc.

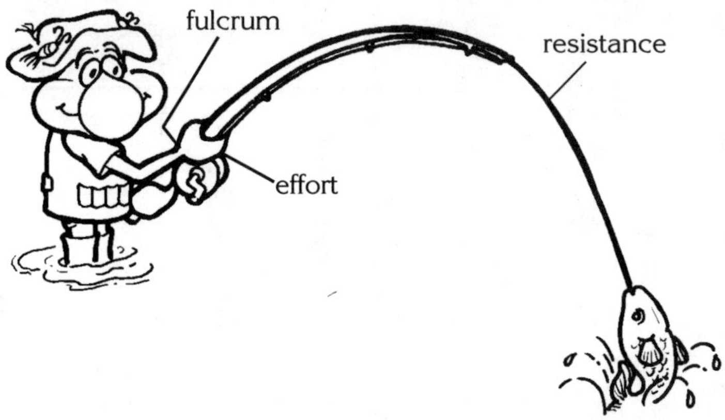

In this activity, students will calculate the actual mechanical advantage that is gained by using a third-class lever. Because the effort of a third-class lever must have a greater force than the resistance, the mechanical advantage will be less than one. The third-class lever exchanges force for distance which is true of all simple machines, but in this case the force is reduced while the distance is increased.

Students are asked to calculate the actual mechanical advantage and the ideal mechanical advantage. They are calculated in similar ways.

$$\text{Actual M.A.} = \frac{\text{Force of Resistance}}{\text{Force of Effort}}$$

$$\text{Ideal M.A.} = \frac{\text{Length of Effort Arm}}{\text{Length of Resistance Arm}}$$

or

$$\text{Ideal M.A.} = \frac{\text{Distance Effort Moves}}{\text{Distance Resistance Moves}}$$

The ideal mechanical advantage is always greater than the actual mechanical advantage due to friction interfering with motion.

Management

1. Allow two 45-minute periods to complete this activity.
2. This activity is best done in groups of four: one to pull the effort, one to measure resistance distance, one to measure the effort distance, and one to measure the stretch of the effort rubber band.
3. The measurements need to be made as accurately as possible (mm) while the lever is held in a steady position. This requires the group to work as a team. It may take some extra instruction to help groups to organize themselves.
4. It may be necessary to experiment with different sizes of rubber bands before using this activity in class. Because of the high degree of elasticity for some rubber bands, the results for the distance portion of this activity may be grossly affected. It is suggested that rubber bands with a lesser amount of elasticity be used.

Procedure

Part 1

1. Discuss the *Key Question.*
2. Distribute the materials to each group.
3. Have students assemble the lever as explained on the instruction sheet.
4. Discuss the meaning of ideal mechanical advantage. Have students calculate the ideal mechanical advantage for the four effort arm lengths by dividing the effort arm length by the resistance arm length. Have students record their solutions.
5. Direct students to hold one ruler vertically behind the meter stick at the first position, 20 cm from the fulcrum. Because the fulcrum is at the 10 cm mark the first position will be at the 30 cm mark which is 20 cm from the fulcrum.Have them note where the top of the meter stick is on the ruler.
6. Instruct one student per group to push down on the meter stick (the effort) at the 30 cm position until the top of the meter stick at the 90 cm position (the resistance) has been lowered four cm.
7. Have the students hold the meter stick in this position while the students designated as the effort measurers measure and record how far the effort has moved from the starting position. Make sure students measure to the nearest millimeter.
8. Repeat steps 5-7 using the same procedure as above but changing the effort position to 40 cm, 60 cm, and 80 cm from the fulcrum.
9. Have the students complete the chart by calculating the distance ratios and their decimal equivalents.
10. Using the data on their charts, direct students to make a line graph of the distance the effort moved and a bar graph of the ideal mechanical advantage.
11. Discuss conclusions students make from the data and graphs.

Part 2

12. Have students transfer the ideal mechanical advantage calculations from *Part 1* onto the chart on *Part 2.*
13. Direct students to measure and record the resting length of the rubber band as it hangs from the meter stick supporting only the paper clip.
14. Have one student place the rubber band at the first position (20 cm from the fulcrum) and pull down the paper clip until the top of the meter stick at the 90 cm position has been lowered 4 cm.
15. With the first student holding the meter stick in this position, have another student measure and record the total stretch of the rubber band.
16. Instruct students to follow steps 14 and 15, changing the effort position to 40 cm, 60 cm, 80 cm from the fulcrum.
17. Direct them to complete the chart by calculating effort stretch and actual mechanical advantage.
18. Using the data on their charts, have students make a line graph of the effort stretch of the rubber band.
19. Discuss conclusions from the data and graphs from both parts of this activity.

Discussion

Part 1

1. This is called a third-class lever. How does it differ from the first- and second-class levers?
2. What is similar about the ideal mechanical advantages for all of the effort arm distances? What does this tell you about this lever arrangement? [The mechanical advantage is always less than one, meaning that with this type of lever, effort is always greater than the resistance]
3. As the distance of the effort arm gets greater, what happens to the distance the effort must move to make the resistance move 4 cm? [It gets greater.]
4. As the distance of the effort arm gets greater, what happens to the ideal mechanical advantage?
5. What similarities and differences are there between the ideal mechanical advantage and the ratio of the distance the forces moved? [They are almost the same value. Differences are due to measurement errors.]
6. How could the measures of the distances the effort and resistance moved be used to determine the mechanical advantage? [Dividing the effort distance by the resistance distance gives you a method of finding the mechanical advantage. The accuracy of the measures affects the accuracy of the solution.]

Part 2

7. As the distance of the effort arm gets greater, what happens to the effort force? [It gets less.]
8. Why are the actual and ideal mechanical advantages different? [Friction reduces actual mechanical advantage.]
9. What comparisons can be made about the graphs? [As the effort moves farther from the fulcrum, the distance moved by the effort and the mechanical advantage get greater, while the effort's force is reduced]
10. What are the advantages and disadvantages of a third-class lever? [The resistance moves a greater distance than the effort, but the effort must be

greater than the resistance giving a mechanical advantage less than one.]

11. Why would someone use a lever like this if it reduces the effort they use? [This lever moves the resistance a long distance while moving the effort a short distance.]
12. What everyday objects are examples of third-class levers? [tennis racket, golf club, fly swatter, fishing rod, paintbrush, broom]
13. Discuss the *Key Question.* [When the effort is closer to the fulcrum than the resistance, the mechanical advantage is less than one. The effort must be greater than the resistance; however, the resistance moves farther than the effort does. When the effort is farther from the fulcrum than the resistance, the mechanical advantage is greater than one. The effort is less than the resistance and the resistance moves less than the effort does.]

Extensions

1. Have students use their graphs to predict what will happen when the effort is applied at different positions on the lever. Have the students test the accuracy of their predictions with the lever.
2. Have students revisit the activity moving the effort one cm at a time and see if their graphs differ.
3. Have students find examples of third-class levers at home and bring them to school. Have students use the examples to explore and see that a small resistance can easily stop the effort's motions and that the effort's small motions move the resistance a great deal.
4. Have students revisit *Clever Lever 2* and determine the ideal and actual mechanical advantages and see how they compare to those of the third-class lever. [The second-class lever will always give a mechanical advantage greater than one and the actual advantage will be less than the ideal.]

Major Conceptual Components

- Energy cannot be created or destroyed.
- The amount of energy or work put into a machine will come out of the machine in some form.
- Energy is used to accomplish work.
- Work is done when something is moved.
- The amount of work done is determined by how much force was used to move the object and how far the object was moved.
- Levers help to lift loads with less effort. The force needed for lifting changes as the distance of the effort arm changes.

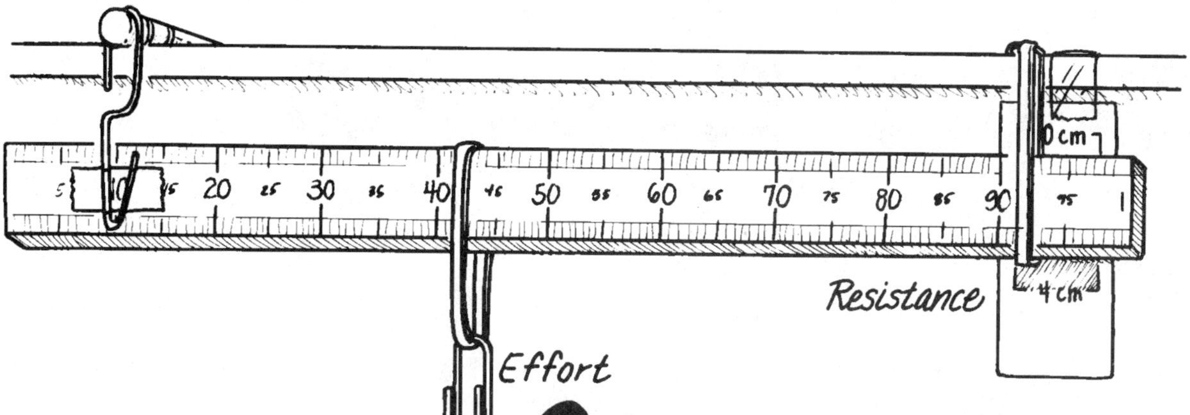

Clever Lever 3

1. Open a paper clip into an S, and tape perpendicular to a meter stick at the 10 cm mark. Tape it so that the larger end of the S sticks out above the top of the meter stick.

2. Tape a pencil to the top of a table so that the end with the eraser sticks out about 2 cm.

3. Hang the meter stick by the end of the paper clip on the end of the pencil. This is the fulcrum of the lever.

4. Slip one rubber band into one of the hooks of each of the two paper clips. Slip both rubber bands onto the end of the meter stick opposite the pencil.

5. Tape the paper clip holding the rubber band nearest the end of the meter stick to the top of the table so the rubber band is at the 90 cm mark of the meter stick. This rubber band will be the resistance on this lever. The other rubber band will represent the effort. It will be used in *Part 2* of this activity.

6. Cut out the marker below and tape it to the table behind the resistance rubber band so that the zero mark is level with the top of the meter stick.

Cut out along dotted line and tape to table behind resistance.

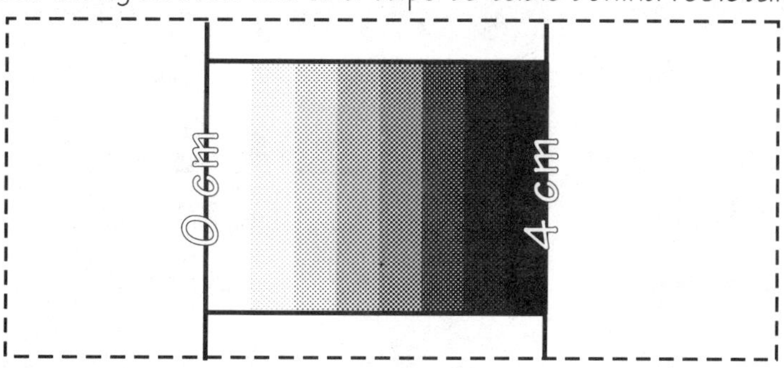

Clever Lever 3

How does a lever work differently when the effort is closer to the fulcrum than the resistance?

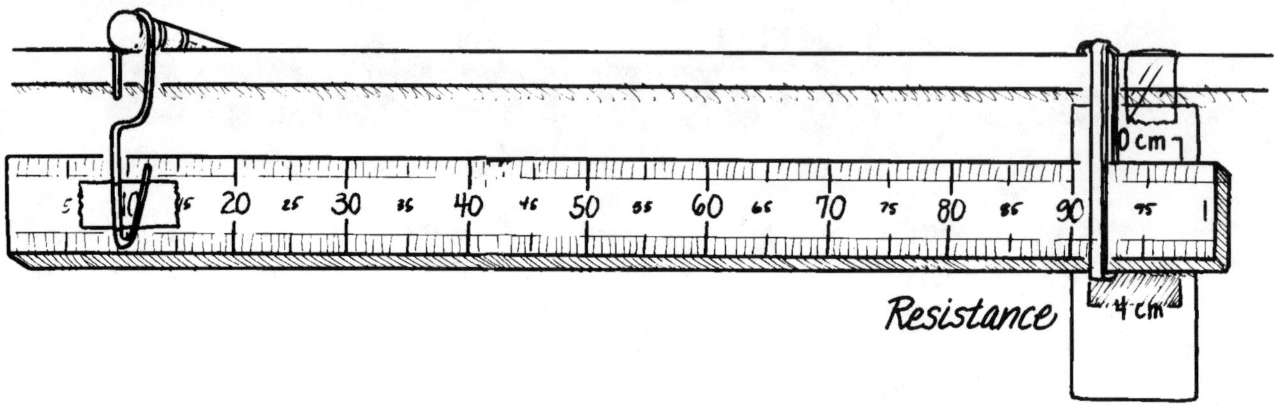

Part 1 - Distance

Push down at each effort arm length until the resistance position at the top of the meter stick has been lowered 4 cm. Measure and record the distance the effort moved the meter stick where it was being pushed. Complete the rest of the chart. Use the data to make a line graph of the distance the effort moves and a bar graph of the ideal mechanical advatage.

Ideal Mechanical Advantage			Distances Forces Move			
Effort Arm Length	Resistance Arm Length	Ideal Mechanical Advantage	Distance Effort Moves	Distance Resistance Moves	Ratio: $\frac{\text{Effort Distance}}{\text{Resistance Distance}}$	Decimal Equivalent
20 cm	80 cm			4 cm		
40 cm	80 cm			4 cm		
60 cm	80 cm			4 cm		
80 cm	80 cm			4 cm		

How could the measures of distances the effort and resistance moved be used to determine the mechanical advantage?

Clever Lever 3

How does a lever work differently when the effort is closer to the fulcrum than the resistance?

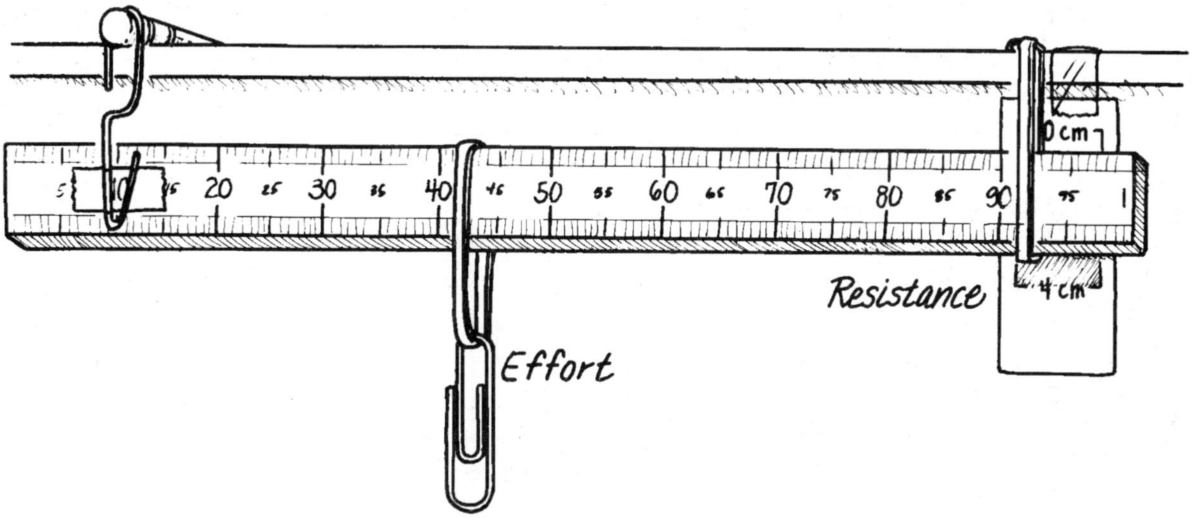

Part 2 - Force

Measure and record the resting length of the effort rubber band when it is supporting only the paper clip. At each effort arm length pull down on the hanging paper clip until the resistance position at the top of the meter stick has been lowered 4 cm. Hold the meter stick in this position to measure and record the total length of the effort rubber band . Complete the rest of the chart. Use the data to make a line graph.

Ideal Mechanical Advantage			Force Applied (Actual Mechanical Advantage)				
Effort Arm Length	Resistance Arm Length	Ideal Mechanical Advantage	Effort Rubber Band: Total Stretch -	Resting Length =	Effort(E) Stretch	Resistance Stretch (R)	Actual Mechanical Advantage E/R
20 cm	80 cm					4 cm	
40 cm	80 cm					4 cm	
60 cm	80 cm					4 cm	
80 cm	80 cm					4 cm	

How do the ideal mechanical advantage and actual mechanical advantage compare?

Clever Lever 3

Distance Graph

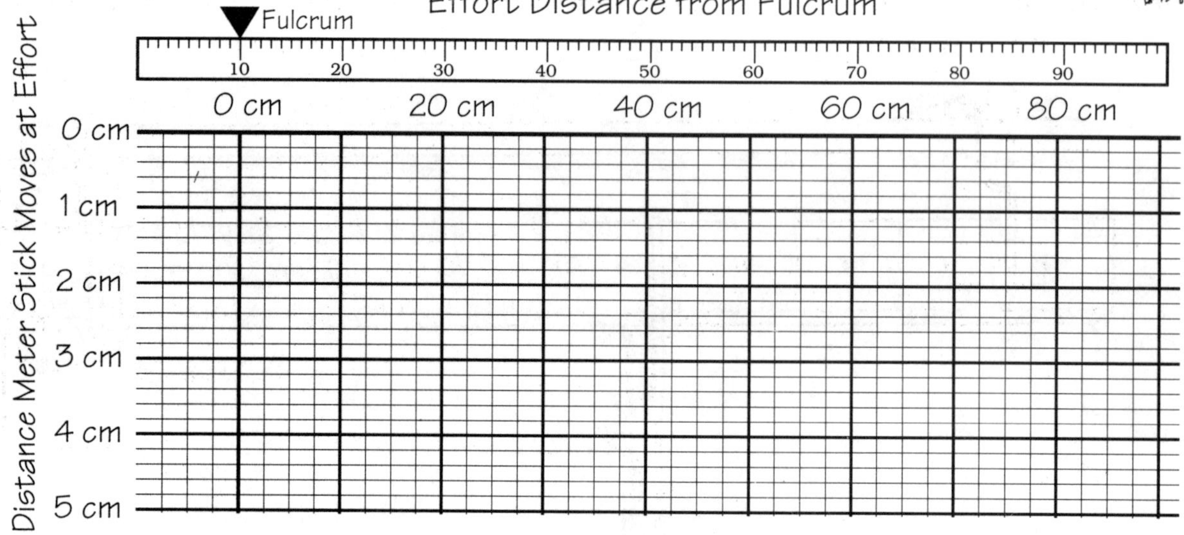

Force Graph

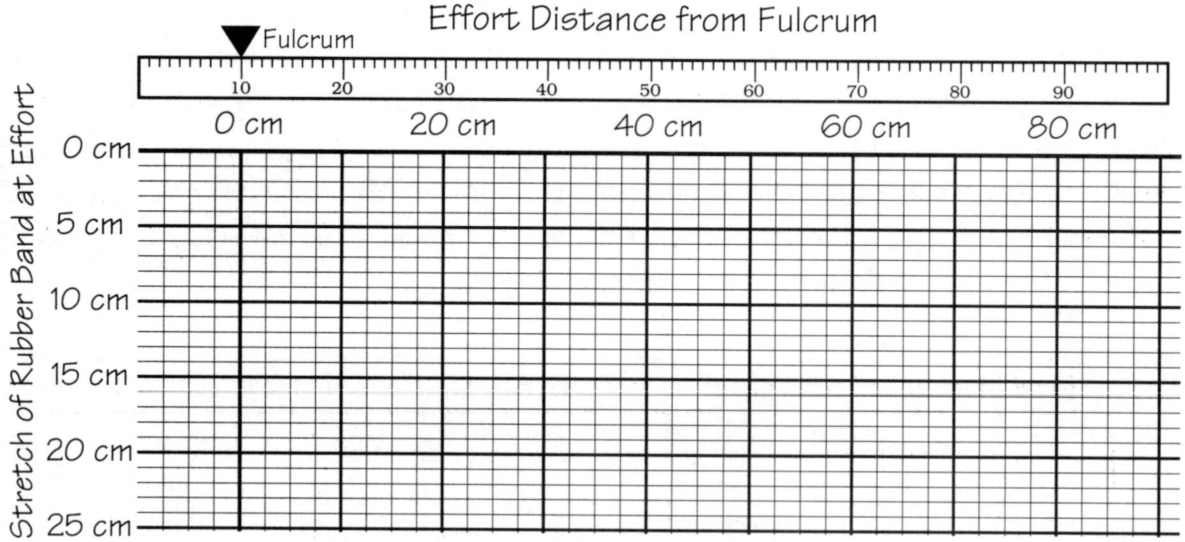

Ideal Mechanical Advantage Graph

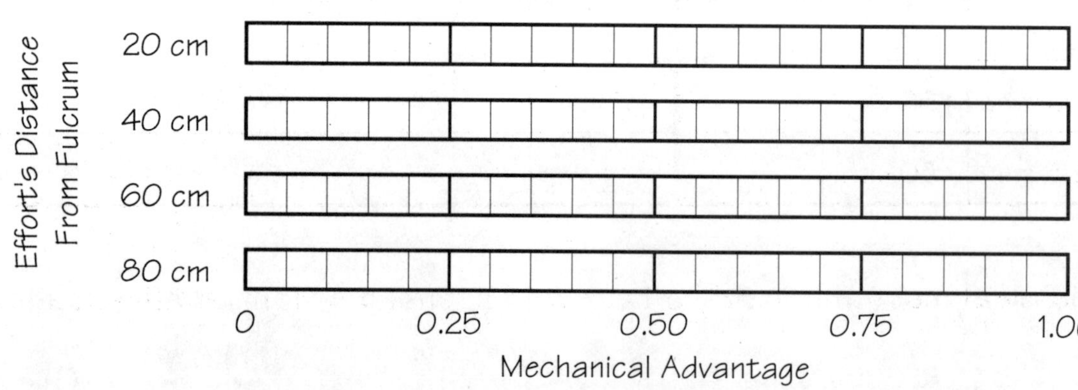

Push-a-Door

Topic Area
Levers and leverage

Introductory Statement
The students will try to devise a way to use a door as an equalizer in a contest between two people who are capable of exerting different forces because of such variables as size and weight.

Math
Measuring
Using rational numbers
- ratios
- proportions
- fractions
- decimals

Using calculators

Science
Physical science
- Simple machines

Math/Science Process Skills
Observing
Making and testing hypotheses
Predicting
Collecting and recording data
Interpreting data
Inferring
Applying

Materials
hinged door
bathroom scale
meter stick
2 500-gram spring scales
masking tape
4 large paper clips

Key Question
You and a friend decide to have a contest to see who can push better. You get on one side of a door and are ready to push when your friend, who is much smaller and on the other side of the door, says it isn't fair. How could you give your friend some advantage to equalize the contest?

Background Information
Second- and third-class levers differ only in the placement of the resistance and the effort forces. In a second-class lever, the resistance is between the fulcrum and the effort force (example: a wheelbarrow). In a third-class lever, the effort is applied to the lever arm between the fulcrum and the resistance (example: a paintbrush).

A hinged door is a lever. The door itself is both the resistance and the effort arms. The hinge is the fulcrum. Because the fulcrum is at one end, the door serves as a second- or third-class lever depending upon where it is pushed (the location of the effort).

Equilibrium refers to the condition when opposing forces acting on a body are balanced. It is a state in which no net force acts. In this activity, when the door is not moving, it is in equilibrium, there is no net torque.

In a state of equilibrium, the amount of force you need to balance an opposing force depends on the forces' respective distances from the fulcrum. To achieve equilibrium with the door, if you push on the door at the doorknob, you only push with half the force of someone who is pushing on the door at the point half the distance from the hinge (fulcrum) to the doorknob. Conversely, the person at the half-way point pushes with twice the force but at one-half the distance. Your push at the knob and the other person's push at the half-way point on the opposite side of the door would balance each other and the door would not move. It would be in equilibrium.

Consider another example: If the doorknob is 4 ft. from the hinge, the half-way point would be 2 ft. from the hinge. The distance ratio is 2:4 or 1/2 (the other person is 1/2 the distance from the fulcrum than you are). The inverse of this ratio is 2/1 which means that the other person must push with two times the force that you do to keep the door in equilibrium.

$$\text{Distance Ratio} = \frac{\text{Partner's Distance}}{\text{Your Distance}} = \frac{2}{4} = \frac{1}{2}$$

$$\text{Force Ratio} = \frac{\text{Partner's Force}}{\text{Your Force}} = \frac{2}{1} = 2$$

The force ratio is the inverse, or reciprocal, of the distance ratio.

Force ratio X Distance ratio = 1 = Equilibrium

Using this example: $\frac{1}{2} \times \frac{2}{1} = 1$

Thus, to be at equilibrium, a person with 1/2 the strength must stand twice the distance from the fulcrum.

Management
1. This activity has two parts and will involve two class periods. The first part involves gaining experience with the concept of inverse ratios. In the second part, the students calculate the "fair game" positions and hold the contest.
2. Before beginning this activity, the teacher must find an appropriate door. It should be one that moves freely on its hinges with no automatic opening/closing mecha-

nism. The ideal door is one which can be positioned perpendicular to the wall to which it is attached. The starting position needs to be located so neither contestant has an advantage. Both contestants should be standing on the same type of surface, and neither should be able to push against a wall or the door jam.

3. In *Part 1* the students will be using spring scales for determining opposing forces. The spring scales need to be pulled perpendicularly to the door to prevent excess friction which may influence the accuracy of the measurement reading. The teacher may want to demonstrate the technique to the class. The activity asks students to perform three trials for each of the distance measures and then to find an average in an attempt to lessen the force measurement error.
4. The students will find their pushing forces by pushing the bathroom scales against a wall. The appropriate unit for measuring force is newtons. Most bathroom scales register weight in pounds or kilograms. The following conversions may be used.

$$(9.8)\ (\text{kilograms}) = \text{newtons}$$
$$(4.45)\ (\text{pounds}) = \text{newtons}$$

Teachers may find it inappropriate for their students to calculate the newtons because of losing understanding of the process through this abstraction. In such cases, students should be encouraged to keep the unit of measure from the scale. Although the unit of force will not be scientifically correct, the calculation can be done adequately with any type of measure of force.

Procedure

Part 1

1. Discuss the *Key Question.*
2. Read the rules for the contest but do not distribute them until *Part 1* is completed.

Rules for Push-a-Door

- The object of the contest is to win the push-a-door game by pushing a door across your win mark on your opponents side of the door.
- Open the door partially as suggested by the teacher and put a mark on the floor under the door. This is the starting mark.
- On each side of the starting mark, 30 centimeters away, put another mark. These two marks are the win marks.
- Opponents stand on opposite sides of a door and are placed so the effect of their forces are equalized.
- The players apply some pressure to their positions while keeping the door at the starting position. On a signal, the players push as hard as they can until the door moves across one of the win lines.

3. Tell students they will be gathering data in *Part 1* of this activity which will be applied to *Part 2*–the Push-a-Door contest.
4. Measure the width of the door from its hinges to the center of its doorknob. Record this on the distance chart as the *distance from the hinge to the stationary scale.*
5. Divide the width of the door into fourths and mark the following locations from the fulcrum with tape: 1/4 (D), 1/2 (C), 3/4 (B), and 1 (A). Point A should be at the center of the doorknob. Record each position's distance from the hinge under *distance from hinge to moving scale* on the distance chart.

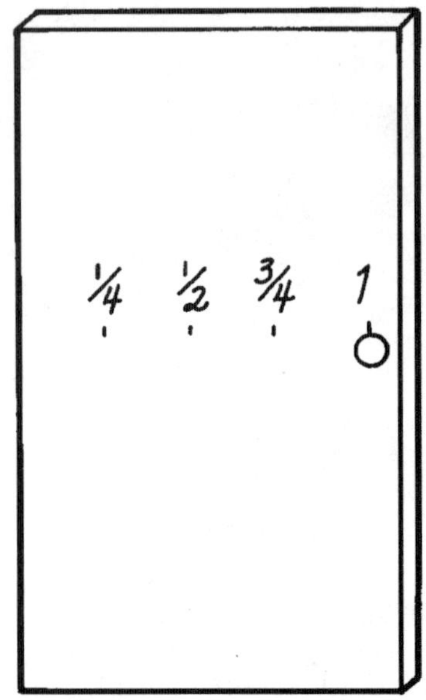

6. Have students write the distance ratios on the chart and calculate their decimal equivalents.
7. Bend the four paper clips into hooks. Tape a hook at each of the four positions. During the activity a spring scale will be attached to these hooks. This scale is referred to as the *moving scale* because its position will vary as it moves from positions A to B to C to D.
8. With the door in an open position, have a student attach a spring scale to the hook at point A while another student, who is on the opposite side of the door, holds onto the stationary scale.
9. The student who is holding the moving scale pulls on it until the door moves and the scale indicates a level of force above the maximum it can record. This student will then reduce the force of the pull until the scale reads four newtons. The door should not be moving at this point. The student holding onto the stationary scale should check its reading. It, too, should read four newtons.The door is now in equilibrium. (If the scales are not both reading four newtons, the students should check to see if they are pulling perpendicularly to the door. Because of the small amount of force being used, students may have to try various strategies for obtaining the best readings.)
10. Students read and record the forces from both scales.
11. Repeat two more times from point A, record the forces, and find the average.

12. Repeat steps six through nine changing the attachment of the moving scale to points B, C, and D.
13. Referring to their record sheets, have the students complete the summary handout.

Part 2

14. Tell students that they are going to apply what they observed and learned from *Part 1*. Ask the *Key Question* again. Distribute the rule sheet.
15. Assign student partners or have students choose an opponent for the contest.
16. Direct participating students to hold the bathroom scale against a wall and push against it to determine their pushing forces. Encourage them to use the same stance for pushing against the scale as they will use when pushing against the door.
17. Have students work through the force ratio and distance measures found at the bottom of the handout which contains the rules for the contest. [The actual number sentence for question five will vary for each set of pushing partners. The general form is: (Decimal Equivalent) x (Weaker Person's Distance from Hinge) = (Strong Person's Distance)]
18. Partners measure and stand at their pushing positions on opposite sides of the door. The opponents begin to apply some force to their positions while maintaining the starting position. A third person says "go" and the opponents push to see who wins the contest.

Discussion

1. How does the simplified ratio on the distance chart compare to the simplified ratio on the force chart? [inverse]
2. How can we use this information to equalize an opponent's force?
3. Why was someone able to win if they took positions that insured fairness and equilibrium? [Answers will vary. Possible factors include friction between shoes and floor, changing pushing force due to endurance, momentum, or changing positions.]
4. A door is a lever. Identify the fulcrum, effort, effort arm, resistance, and resistance arm in the contest. [fulcrum: hinge; effort: first opponent; effort arm: distance from hinge to first opponent; resistance: second opponent; resistance arm: distance from hinge to second opponent]
5. Explain why you think a door is a second- or third-class lever. [For the weaker person, the door was a second-class lever because the resistance was between the fulcrum and the effort. For the stronger person, the door was a third-class lever because the effort was between the resistance and the fulcrum.]

Extensions

1. Have students determine a system to adjust the positions, rules, or "playing field" to equalize the opponents. Have them use their system to see if the contest is better equalized.
2. Have students research ways athletics compensate for the skills and forces in athletes. [weight classes, handicaps, difficulty ratings]

Major Conceptual Components

- Energy cannot be created or destroyed.
- The amount of energy or work put into a machine will come out of the machine in some form.
- Energy is used to accomplish work.
- Work is done when something is moved.
- The amount of work done is determined by how much force was used to move the object and how far the object was moved.
- Levers help to lift loads with less effort. The force needed for lifting changes as the distance of the effort arm changes.

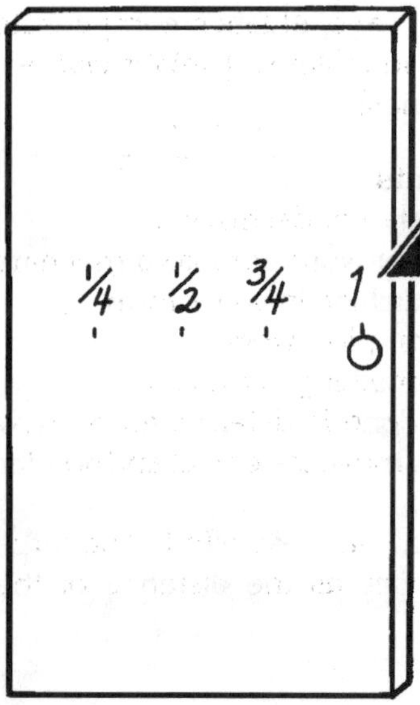

PUSH -A- DOOR

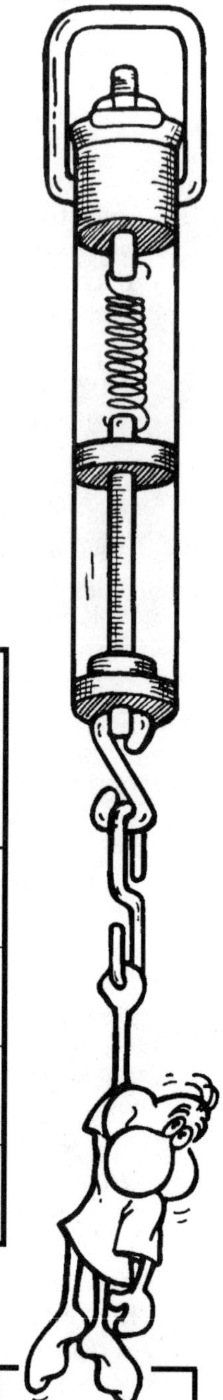

Distance Chart

Position of Moving Scale	Distance from Hinge to Moving Scale	Distance from Hinge to Stationary Scale	Distance Ratio: $\frac{\text{Distance to Moving Scale}}{\text{Distance to Stationary Scale}}$	Decimal Equivalent
A (1)				
B (3/4)				
C (1/2)				
D (1/4)				

Force Chart

Position of Moving Scale	Force on Moving Scale (N)				Force on Stationary Scale (N)				Force Ratio: $\frac{\text{Stationary Force}}{\text{Moving Force}}$	Decimal Equivalent
	Trial 1	Trial 2	Trial 3	Average	Trial 1	Trial 2	Trial 3	Average		
A (1)										
B (3/4)										
C (1/2)										
D (1/4)										

PUSH-A-DOOR

Use the information from the Record Sheet to complete:

For Position A:

In the Distance Ratio the distance of the ________________ scale is the same as the distance of the ________________ scale.

In the Force Ratio the ________________ scale exerts the same force as the ________________ scale.

For Position B:

In the Distance Ratio the distance of the ________________ scale is 3/4 the distance of the ________________ scale.

In the Force Ratio the ________________ scale exerts 3/4 the force of the ________________ scale.

For Position C:

In the Distance Ratio the distance of the ________________ scale is _________ the distance of the ________________ scale.

In the Force Ratio the ________________ scale exerts _________ the force of the ________________ scale.

For Position D:

In the Distance Ratio the ________________ scale is the ________distance of the ________________ scale.

In the Force Ratio the ________________ scale exerts _________ the force of the ________________ scale.

For Position N:

In the Distance Ratio the distance of the moving scale is 1/n the distance of the stationary scale.

Now write the sentence that goes with the above statement.

PUSH-A-DOOR

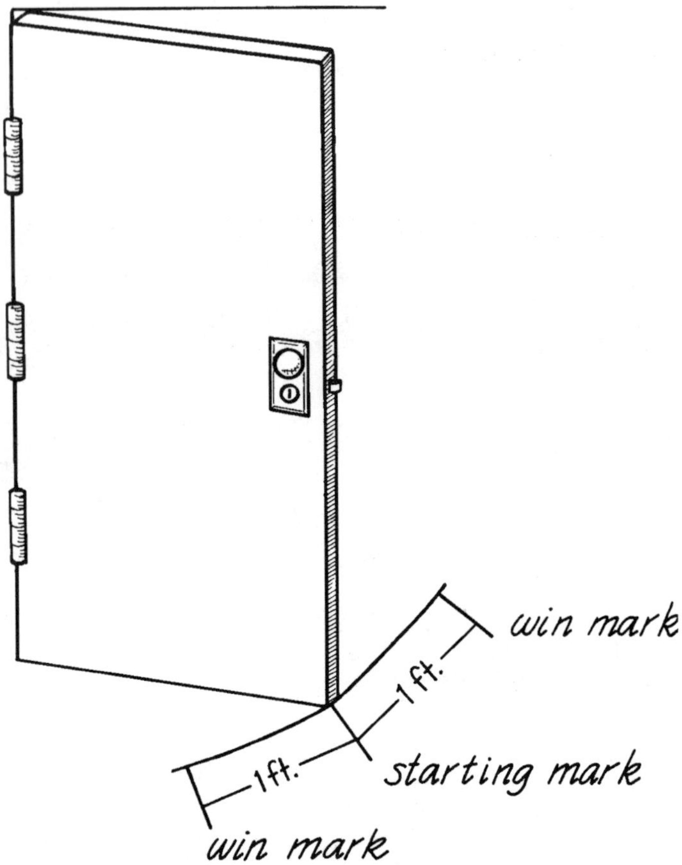

Rules of Push-A-Door Contest

A. Object of game: Win a push-a-door contest by pushing a door across the win line on the opponent's side of the door.

B. Mark starting position and 2 win positions on the floor. (Use a temporary mark such as chalk.)

C. Opponents stand on opposite sides of the door in pre-calculated positions so the effects of their forces are equalized.

D. On a signal, the players push on the door. The first to force the door over the opponent's win mark wins the contest.

1. Pick an opponent. Each of you push a scale against a wall to measure your pushing force.

My Force = ________ Opponent's Force = ________

2. Calculate the force ratio.

$$\frac{\text{Force of Weaker Person}}{\text{Force of Stronger Person}} = \frac{\square}{\square} = \square$$

(Force Ratio) (Decimal Equivalent)

3. You now know the force ratio. In the force ratio the weaker person has _______ the force of the stronger person. Therefore, in the distance ratio the stronger person should stand _______of the distance of the weaker person.

4. The weaker person will stand at the doorknob, which is _______ cm from the hinge. So, the stronger person should stand _______ of _______ cm for a fair contest.

5. Write number 4 as a number sentence (algebraic expression).

Lever Assessment Pages

The following two lever activities have been included as reinforcement or assessment experiences.

The activities may be presented as group or individual activities at the discretion of the teacher. Both activities are designed as cut and paste activities where student use what they have discovered about levers to sort them.

Class Acts, the simpler of the two activities, asks students to identify the effort, resistance, and fulcrum of the pictured levers. They will then sort the pictures into the three classes of levers and discuss the characteristics common to each class.

The three classes are listed below.

1st Class
can opener
seesaw

2nd Class
wheelbarrow
bottle opener
nut cracker

3rd Class
broom
fly swatter
hammer
forearm

Sorting Levers has students sort the levers by mechanical advantage. Students are asked to calculate the mechanical advantage and classify each lever. An answer key is included for the teacher's reference.

Class of Lever	Example Levers	Mechanical Advantage
2nd	 E	6:1 = 6/1 = 6
1st	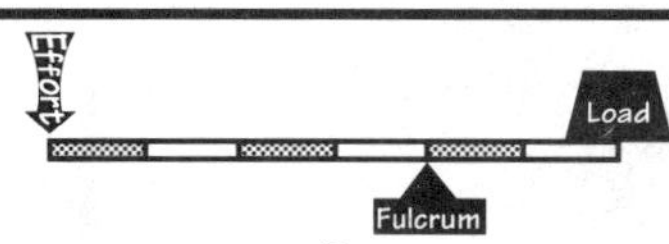 B	4:2 = 2/1 = 2
2nd	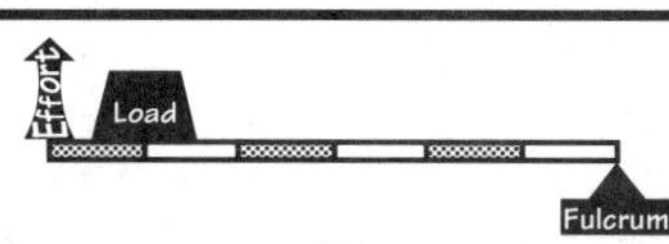 F	6:5= 6/5 =1.2
1st	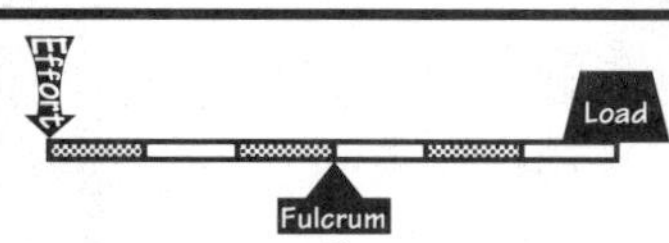 G	3:3 = 3/3 = 1
3rd	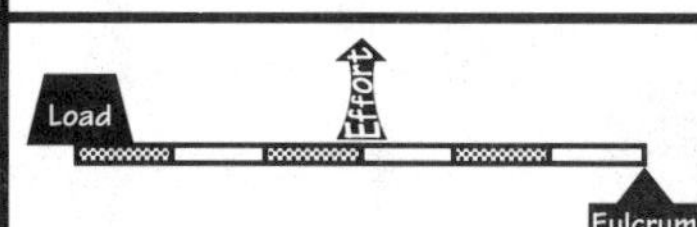 A	3:6 = 1/2 = .5
1st	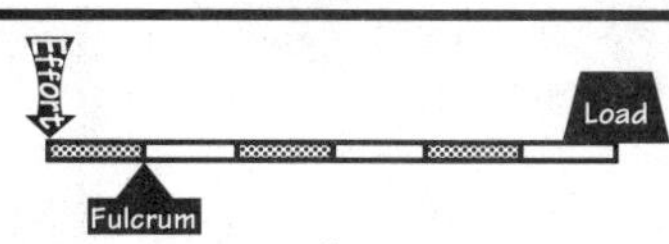 D	1:5 = 1/5 = .2
2nd	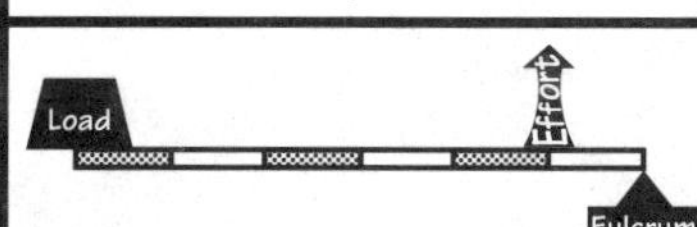 C	1:6= 1/6 = 0.16 2/3

CLASS ACTS

1. Cut out the nine examples of levers in action below.
2. On each example label the fulcrum, the resistance that is being moved, and where the effort is applied to get the work done. You may need to complete some drawings.
3. Sort the levers into the three classes of levers.
4. On a separate piece of paper, paste down the levers in the three classes and label each group.
5. Write a summary statement that tells how all the levers in each class are similar.

SORTING LEVERS

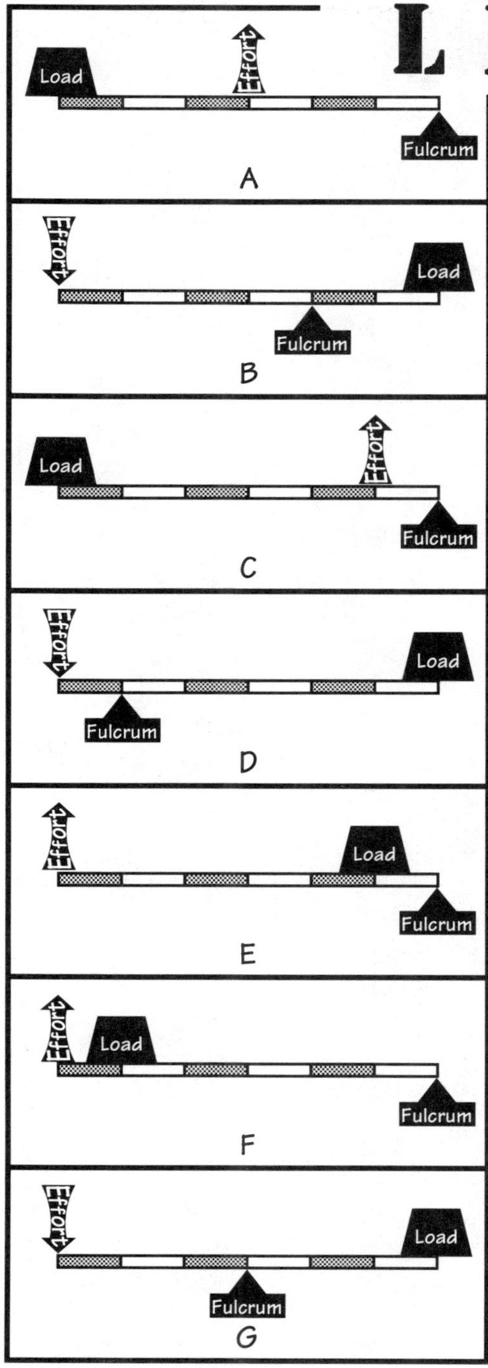

1. Cut out the seven lever examples to the left.
2. Sort them until they are in order from the one with the greatest mechanical advantage to the one with the least mechanical advantage.
3. *Glue each of the levers down within the dashed lines to the right. Keep them in order from greatest to least.*
4. *For each lever example calculate the mechanical advantage in the region to the right of the example.*
5. Determine what class of lever is shown in each example and record in the region to the left.

Class of Lever	Example Levers	Mechanical Advantage
	Greatest	
	Least	

October 16

Dear Journal,

There has been a slow steady rain all day so Mucky and I played in the shop. Now that we understand how to assemble the levers, we put some together and started messing around with them. We even came up with a game. Mucky would put a block on one end of a lever, and then I had to drop a different block onto the other side of the lever. The trick was to get Mucky's block to be thrown up high enough to land on top of the work bench. It took a lot of trials to get it to work just right.

It was during this game that Handy walked in to see what was keeping us busy. He said that catapults were fun but that there was a much easier way to determine the height from which I should drop my block. He then pulled out the encyclopedia and showed us this paragraph:

> In an ideal system the potential energy going into the system is equal to the kinetic energy released by the system. This relationship can be expressed in the equation: $mgh = mgh$.

Seeing our puzzled looks, Handy went on to explain. Energy is either being stored or it is being used. Energy that is being stored is called **potential energy**. A stretched rubber band in a slingshot has potential energy. The block of wood you hold up in your hand has potential energy. When that potential energy is released, something moves. Energy of motion is **kinetic energy**. When you release the rubberband of the slingshot, the rock shoots out. The shooting rock has kinetic energy. When you let go of the block, it hits the catapult and makes the other block move. This is kinetic energy.

The equation says that the amount of potential energy I put in is the same as the amount of kinetic energy I get out. Handy then used our catapult to demonstrate. He put two blocks on the catapult. Next, he measured the height of the work bench which was 1 meter. Then, he showed us he was going to drop 3 blocks at the same time onto the other end of the catapult. He pulled out a piece of paper and wrote this:

Potential Energy	=	Kinetic Energy
mgh	=	mgh
mass x gravity x height	=	mass x gravity x height
3 x 9.8 m/sec^2 x ?	=	2 x 9.8 m/sec^2 x 1
3 x ?	=	2 x 1
3?	=	2
3?/3	=	2/3
?	=	2/3
Drop height	=	2/3 m

Handy said that gravity is the same on both sides of the equation so we can ignore it. That means if I know the mass of Mucky's block, and how high we want it to go, the mass of my block times the height I drop it from has to equal Mucky's block's mass x height.

Handy dropped the three blocks from 67 cm. Amazingly, the other blocks flew up and just reached the bench. It was so awesome because Handy did it on the very first try!

We had a great time playing with the catapults the rest of the afternoon.

October 21

Dear Journal,

We are definitely having the last of a beautiful Indian summer. The cold, rainy weather is gone and we are sweating it out again. Mucky is complaining the most with that awful cast. His leg sweats and itches something awful. He says that when he gets the cast off on Saturday, he is heading straight for the swimming pool for a refreshing plunge.

October 23

Dear Journal,

Well, Mucky went to the pool to take his plunge today. He sprawled out to do one of his famous belly-flops. Too bad he didn't check the water first. I found him lying at the bottom of the pool in the clutches of Greedy Gravity. Luckily his natural padding kept him from any serious injury. When I got there, he was lying on his back pleading for me to help him out, saying he was in pain and could not move.

What was I to do this time? I had to lift Mucky 2 meters to the top of the pool deck. He has a mass of 300 kilograms and I cannot lift that much. It occurred to me that I had to use a machine to help me out. It was then that I spotted the 3 meter diving board and recognized that I could make a catapult. I could push something off the board and catapult Mucky up onto the deck.

I raced to the shop to get the materials. I wanted to be a little more careful this time because I felt kind of guilty about the painful mess I caused when I greased Mucky to reduce the forces of Fiendish Friction and he plowed into the car breaking his ankle. I didn't want to send him on another trip to the hospital, so I sat down and did some thinking. Here's what I came up with:

Resistance (that's Mucky) = 300 kg
Height out of Pool = 2 m
Height of Drop for board = 5 m
Mass of dropped object = ?

mh = mh (I left gravity out because I remembered it was the same for both sides of the equation).

I did my calculations and raced back to the pool with my supplies to help Mucky. I'm really glad I listened to Handy because the catapult worked on the first try. Imagine if it hadn't!! Ooh, that's a painful thought!

Editor's Note: Fantastic Force's calculations seemed to have gotten lost in the rush. Can you reproduce them and find the mass that was dropped from the information?

May The Force Be With You

Topic Area
Force, energy, and energy conservation

Introductory Statement
In this group of investigations (*May The Force Be With You, May The Force Be With You II,* and *Return Of The Force*), the students will be isolating variables in order to learn about the energy balance that is nature.

Math
Averaging
Measuring
- length
- mass

Using formulae
Using rational numbers
- ratios
- percentages

Graphing
Using calculators

Science
Physical science
- Simple machines

Math/Science Process Skills
Observing
Predicting
Identifying and controlling variables
Measuring and comparing
Collecting and recording data
Interpreting data
Inferring

Materials
Per group of five students:
- 2 1/4" diameter X 35" long dowels
- 2 inch ball of clay
- 1 50 cm length of 14 gauge galvanized wire (for catapult)
- 1 25 cm length of 14 gauge galvanized wire (for platform)
- 10-15 wide rim fender washers (washers with small holes but large total diameter) $^1/_4$" hole, $1^1/_4$"diameter
- 2 10 cm X 25 cm pieces of pegboard with $^1/_4$" diameter holes
- pliers
- balance
- masses–70 grams
- 1 small empty thread spool
- meter stick
- chart paper, chalkboard, or overhead projector

Key Question
What variables affect how high an object will go off the end of a catapult?

Background Information
The First Law of Thermodynamics states that there is an energy balance in nature: **energy in = energy out**.

The potential energy that an object has is its mass x gravity x height (mgh). Mgh is a measure or a unit of energy. In the catapult lessons that follow, the students will do a variety of investigations which will lead them to an understanding of mgh. On one side of the catapult, we have the mass (the effort–m) under the force of the gravitational pull of the earth (g) being dropped from a certain height (h). By multiplying these three quantities, we have mgh or potential energy on the effort side of the catapult.

Once the mass is released, the following chain of events takes place in what is called an energy balance:

effort arm **resistance arm**

potential > potential-kinetic > kinetic > kinetic-potential > potential

As the mass is released, the potential energy changes into kinetic energy (the energy of motion) causing the object on the resistance arm (the resistance) to be forced to a certain height on the other side of the catapult.

By manipulating the variables of mass and height, the students will begin to recognize the energy balance in nature. They will discover that the energy going in (mgh of dropped mass) is approximately equal to the energy going out (mgh of object catapulted). For example: If a 40 gram weight is dropped from a 45 cm height, how high do you have to drop a 50 gram weight in order to force the resistance to reach approximately the same height as with the 40 gram weight? In this case, you know the mass and the height of the effort and you want the height of resistance to remain the same when you are using a different mass. You can calculate mgh = mgh. Because the force of gravity remains the same (9.82 cm/sec^2) whether it is on the potential energy side or the kinetic energy side of the catapult, it cancels itself out. Thus, we are left with mh = mh.

$$\begin{aligned} 40 \text{ g} \times 45 \text{ cm} &= 50 \text{ g} \times ? \\ 40 \times 45 &= 50 \times \text{height} \\ 1800 &= 50h \\ 1800/50 &= h \\ 36 &= h \end{aligned}$$

Therefore, a 50 g weight dropped from 36 cm and a 40 g weight dropped from a 45 cm height will force the resistance to reach approximately the same height. Some error in results will be caused by friction from the effort or the resistance; the fulcrum, not being at the midpoint; leaning dowels; unevenly distributed weights; etc.

Consider the journal story of Mucky Mass. He weighs 300 kg and needs to be lifted six meters. The mass will be dropped from the 3 meter board for a total drop of nine meters. Thus,

$$\begin{aligned} ?\text{ kg} \times 9\text{ m} &= 300\text{ kg} \times 6\text{ m} \\ h \times 9 &= 300 \times 6 \\ 9h &= 1800 \\ h &= 1800/9 \\ h &= 200 \end{aligned}$$

Fantastic Force will need to drop at least 200 kg of mass from the diving board.

Management

1. This activity is divided into three parts. In *May The Force Be With You*, students isolate the variable of effort's drop height. In *May The Force Be With You II*, the students experiment with the effort's mass. *Return Of The Force* has students combine what they know about the effort's mass and height to get the resistance to go a given height.
2. In preparation for teaching the activities, make a catapult for your own experimentation and to use as a model for the students to follow as they build their own.
3. The activities work well with students in groups of five.
4. To complete the set of activities requires four to five periods: one period to construct the catapult system, one period for *May The Force Be With You*, one period for *May The Force Be With You II*, a fourth period for *Part 1 Return Of The Force*, and another period for *Part 2 Return Of The Force.*
5. Use 14 gauge wire so the catapult will be sturdy enough to withstand the force of the weights.
6. When making the weights, use fender washers and a minimal amount of clay. To economize on the number of washers necessary for these activities, have the students keep adding washers to make heavier weights (Ex. 30 g = 3 washers, 40 g = 4 washers, etc.).
7. Careful construction of catapults is necessary for the appropriate outcomes of the activities. Stress that the students build the catapults so they are balanced. Have the students test and adjust their catapults so friction of the masses and the spool are minimal. Be aware that final outcomes are prone to some error due to the simple crudeness of the systems.
8. To culminate the series of activities on levers, the teacher may want to consider *The Great Catapult Caper* which ties together the elements of these force activities and the lever lessons.

Procedure

Construction of the catapults:

1. Divide students into groups of five. Distribute the handout *Catapult Construction.*
2. Have students construct the catapults and mark the dowels with lines indicating 5 cm intervals. Set up the structure using the two pieces of pegboard as a base.
3. Brainstorm with the class the variables which they believe will affect the height the resistance will go. Write the suggested variables on the chalkboard, chart paper, or the overhead projector to be used later for a discussion of the effects of the individual variables.

Activity 1: May The Force Be With You

1. Discuss the questions: If we drop the same mass (30 g) from different heights onto a catapult, how will it affect the height the resistance (spool) rises? How could you predict how high the spool will go before you dropped the mass?
2. Distribute and review the activity sheet *May The Force Be With You.* Explain the procedures of making the weights.
3. Caution students to make certain that they need to hold the dowels straight during the trials.
4. Have the students follow the directions, observe and record the results, and complete the graph.
5. As students complete the activity, discuss what relationship exists between the height the effort was dropped and the height the resistance traveled? [The height of the resistance increased as the height of the drop increased.]

Activity 2: May The Force Be With You II

1. Ask: What will be the effect on the spool (resistance) when we change the weight but not the height of the effort in our catapult system?
2. Distribute and review the activity sheet *May The Force Be With You II.*
3. Suggest to students that they can use the same washers for the various weights by adding clay and more washers.
4. Have them conduct the trials, then record, average, and graph their results.
5. At the completion of the activity, discuss the following questions: What effect did changing the weight have on the height the spool (resistance) rose? [The heavier weight forced the resistance higher] Does increasing the weight dropped have the same effect as increasing the height the mass is dropped? Explain.

Activity 3: Return Of The Force
Part 1

1. Discuss how you could determine from what height to drop a weight to get the spool to rise a given distance.
2. Using the catapult equipment, have students take a 30 g weight and drop it so that it causes the spool to rise to the 40 cm mark on the dowels.
3. When the drop consistently gives a 40 cm rise, direct the students to record the height of the drop.
4. Have students calculate the mass x height on the effort arm data.
5. Repeat steps two to four changing the height of the resistance to 60 cm as indicated on the activity page.
6. Have students increase the weight of the washers to

40 g, 50 g, 60 g, and 70 g repeating steps two to five for each different mass.

7. Have students discuss what pattern they see in the mass x height of the effort's column. [To get the resistance to rise to a given height, the mass times height of the effort is always approximately the same.]

Part 2

1. Have students experiment with the height of the drop needed to get a 30 g weight to raise the resistance to 50 cm.
2. Direct the students to record the drop height needed to make the given rise and calculate the mass times height in the first row on the chart.
3. Using the mass times the height from the top row of the chart, have the students predict what height drop will be needed for a 40 g, 50 g, and 60 g weight.
4. Allow time for the students to try the different masses to see how close their predictions are to the actual outcomes and record their results on the chart.
5. Have students discuss their conclusions and write a summary statement and formula.

Discussion

1. What is the relationship between the mass and the height?
2. What happens to the height of the effort as you increase the mass of the effort when holding the resistance constant?
3. What is the energy balance that exists?

Curriculum Correlation

History:

Have students research the history of the catapult including the significance of engineers to the war effort (especially Archimedes). The reports can be done individually or with a group, oral or written.

Major Conceptual Components

- Energy comes in many forms.
- Mechanical energy is the energy due to the position or the movement of something.
- Energy cannot be created or destroyed.
- The amount of energy or work put into a machine will come out of the machine in some form.
- Energy is used to accomplish work.
- Work is done when something is moved.
- The amount of work done is determined by how much force was used to move the object and how far the object was moved.
- Levers help to lift loads with less effort. The force needed for lifting changes as the distance of the effort arm changes.

CATAPULT CONSTRUCTION

1. CATAPULT:
Bend a 50 cm wire into a U shape. Use pliers to make corners.

2. Twist the midpoint of the wires around a dowel one at a time to form circles.

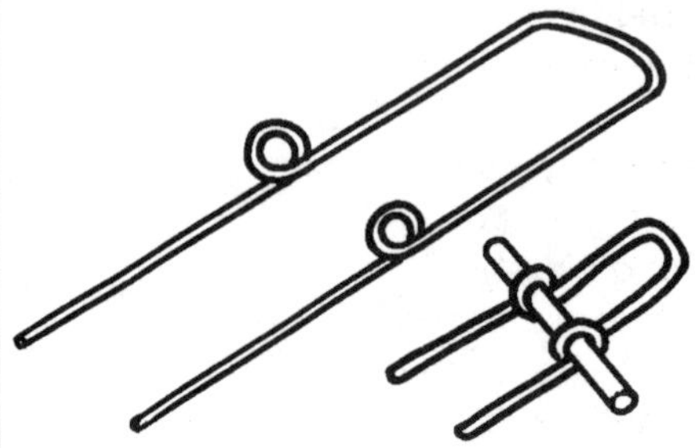

3. Close off the end by bending one side to form a corner and wrap it around the other end.

4. PLATFORM:
Find the midpoint of a 25 cm length of wire. Use pliers to form the M shape shown. Bend up the ends so that it will stand.

5. Thread the catapult through the platform so that it hangs down under the platform.

6. Use a meter stick. Mark dowels at 5 cm intervals.

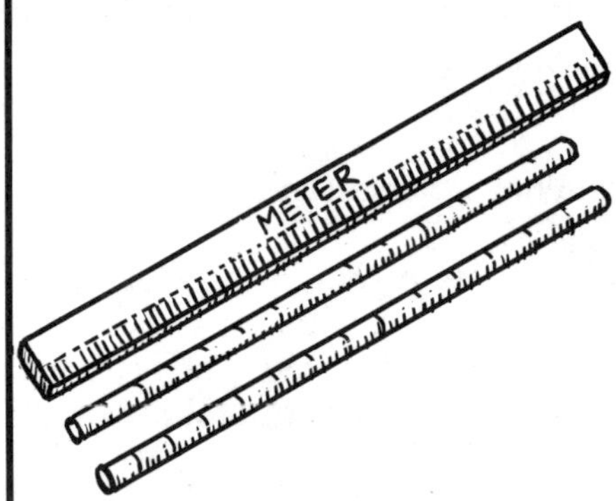

7. Place a dowel at either end of the catapult. Secure dowels in a pegboard base. Use clay to secure platform to the base.

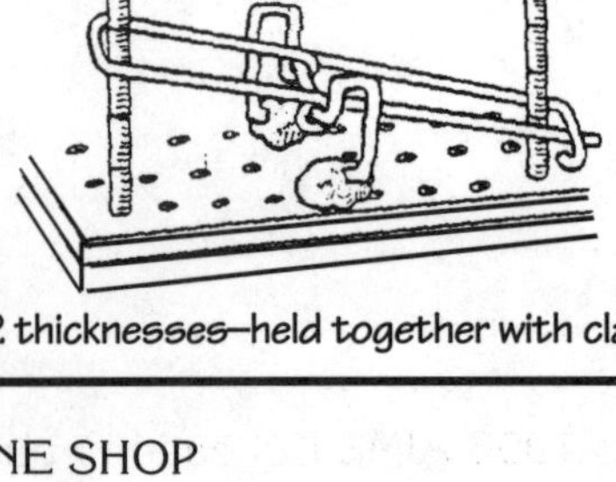

2 thicknesses—held together with clay.

8. Make weights according to the amount needed. Sandwich dots of clay evenly between washers. Increase the weight by adding more washers. Continue to sandwich with clay.

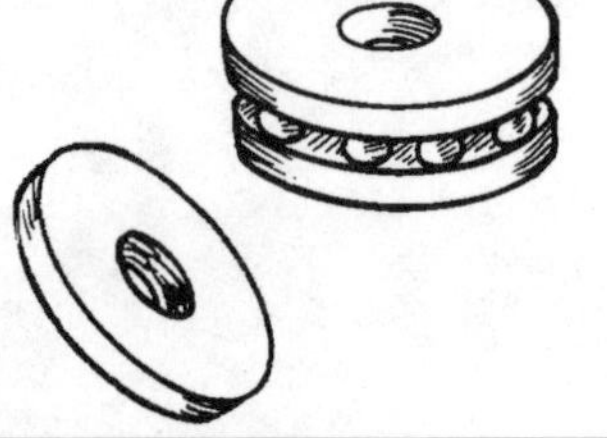

9. Place spool over one end of a dowel. Washer weight goes over the other dowel.

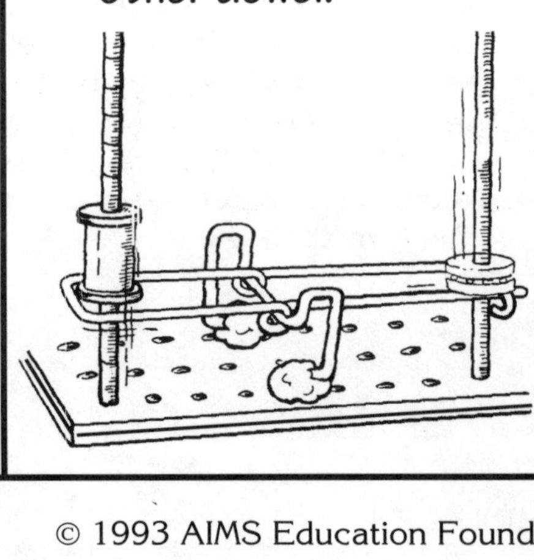

MAY THE FORCE BE WITH YOU

What will be the effect on the resistance when the same mass is dropped from different heights?

Controlled Variable: ____________ Manipulated Variables: ____________

1. Construct the catapult and platform.
2. Make a 30 gram weight with washers and clay.
3. Drop the weight (effort) from the heights listed, 3 times each.
4. Measure the height the spool (resistance) reaches each time. Find the average.
5. Make a graph to display the results.

Effort Height of Drop (cm)	Height of Resistance (cm)			
	Trial 1	Trial 2	Trial 3	Average
80				
70				
60				
50				
40				
30				

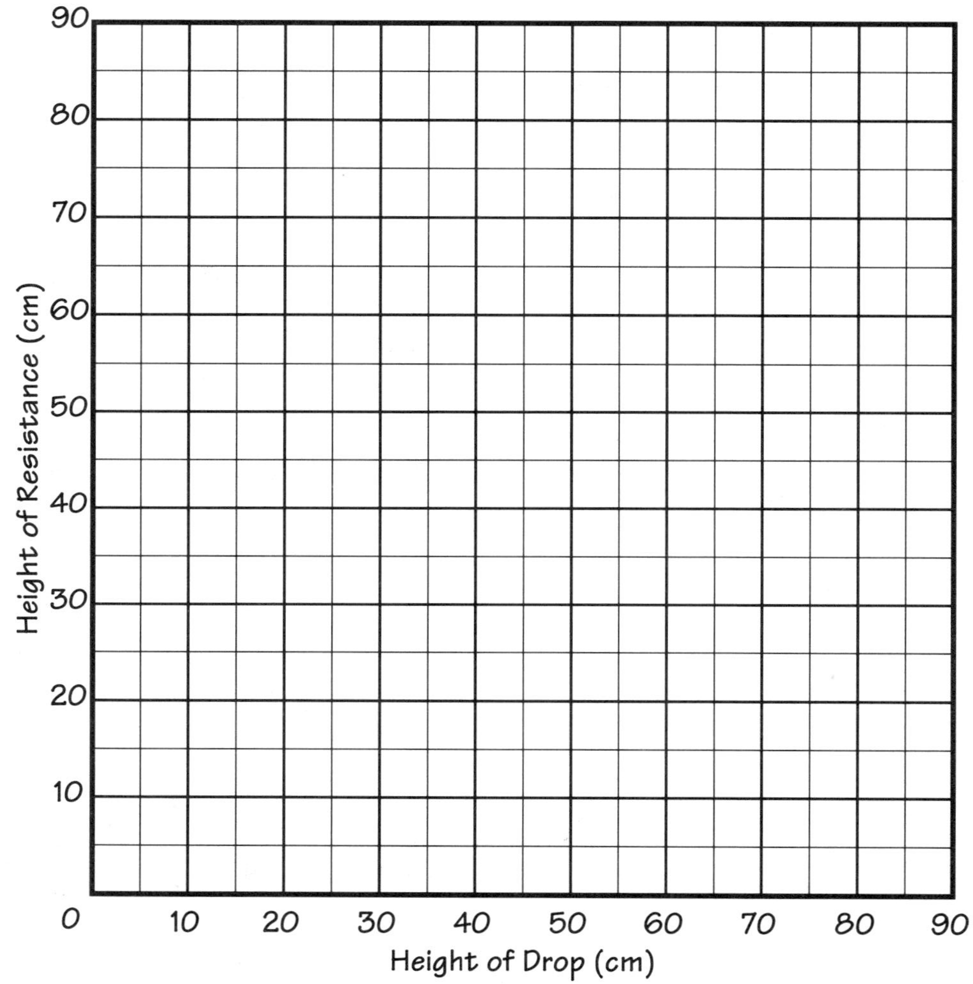

MAY THE FORCE BE WITH YOU II

What will be the effect on the resistance when we change the mass but not the height of the effort?

Controlled Variable: ______________

Manipulated Variables: ______________

1. Set up the catapult and platform.
2. Make a 30 gram weight and then drop it from a 50 cm height 3 times. Find the average.
3. Add weight to the effort making 40g, 50g, and 60g. Repeat step 2 for each new weight.
4. Graph the results.

Effort Mass (g)	Height of Resistance (cm)			
	Trial 1	Trial 2	Trial 3	Average
30				
40				
50				
60				

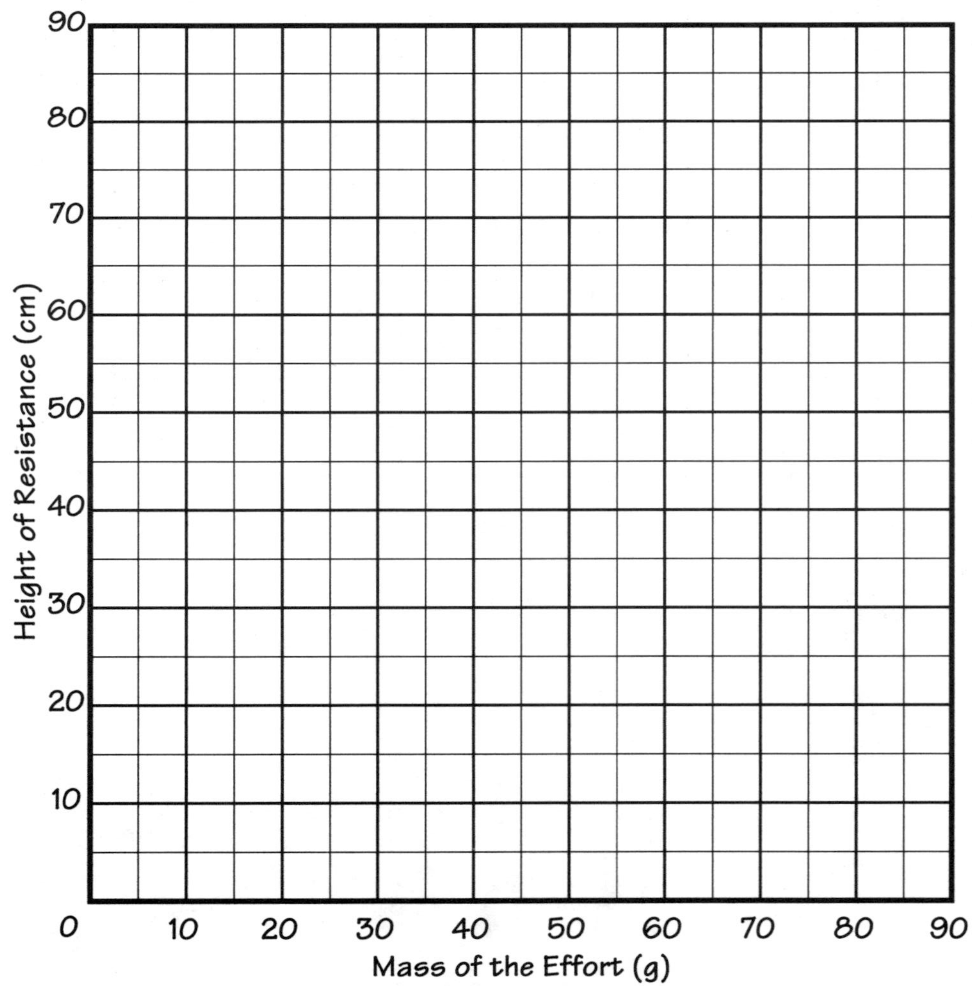

Return of the Force

Hold the height of the resistance constant, change the mass and experiment to find the necessary height from which to drop the effort.

Controlled Variable: ______________ Manipulated Variables: ______________

Height of Resistance (cm)	Mass of Effort (g)	Height of Effort (cm)	Mass X Height ($m_e \times h_e$)
40	30		
40	40		
40	50		
40	60		
40	70		

Height of Resistance (cm)	Mass of Effort (g)	Height of Effort (cm)	Mass X Height ($m_e \times h_e$)
60 cm	30 g		
60 cm	40 g		
60 cm	50 g		
60 cm	60 g		
60 cm	70 g		

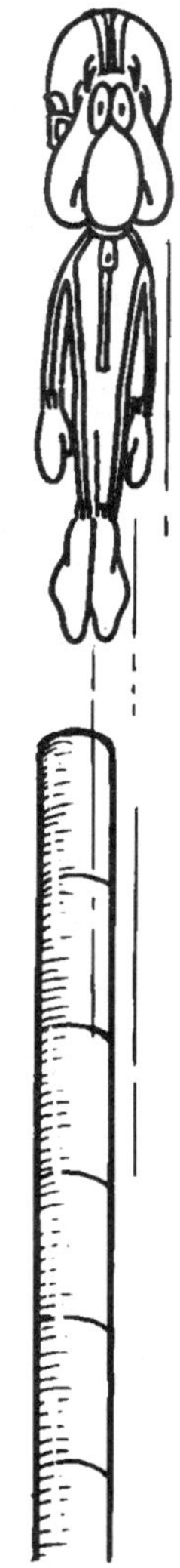

What is the pattern in mass x height of effort? ______________

Return of the Force

Part 2

Predict the height that would be necessary to drop the effort when the resistance is held constant.

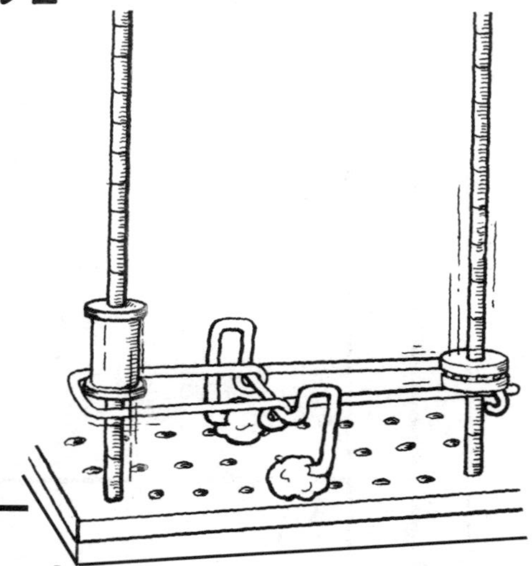

1. Experiment to find height in the first case. (30 g)
2. Predict height in the other cases. (40 g, 50 g, 60 g)
3. Then find the actual heights for each case.
4. Perform all calculations. Find your percent of error.

Predicted Height of Effort (cm)	Predicted $m_e \times h_e$	Height of Resistance (cm)	Mass of Effort (g)	Actual Height of Effort (cm)	Actual $m_e \times h_e$	Percent of Error (%)
		50	30			
		50	40			
		50	50			
		50				

5. Write a formula to show what you have learned: ____________________

The Great Catapult Caper

Topic Area
Levers

Introductory Statement
Students will apply the information they have learned about catapults and levers in order to build a high performance catapult.

Math
Averaging
Measuring
Using formulae
Using rational numbers
  ratios
Graphing
Using calculators

Science
Physical science
  Simple machines

Math/Science Process Skills
Observing
Predicting
Measuring
Comparing
Collecting and recording data
Inferring
Identifying and controlling variables

Materials
See *Materials* and *Catapult Construction* from *May The Force Be With You*

Key Question
How will you build a catapult to win a high performance catapult contest?

Teacher Background

A lever is a bar that rests on a fixed point called the fulcrum. The force exerted on the lever is called the effort. The object that the lever lifts is called the resistance. The distance from the fulcrum to the point where the effort is exerted is called the effort arm. The distance from the fulcrum to the point where the resistance is located is called the resistance arm. The closer the fulcrum is to the resistance, the less effort will be needed to move or lift the resistance. However, this means that the effort must move over a longer distance while the resistance moves a shorter distance. When the opposite occurs, the effort arm is closer to the fulcrum than the resistance arm, a greater force will be needed to lift the resistance. In this case, the effort will move a shorter distance while the resistance moves a greater distance.

Management
1. This activity should follow the other lever activities and the series of activities found in *May The Force Be With You*.
2. This activity can be assigned as homework and brought back to class for the competition, or it could be assigned as class work with students working in groups to design their catapults.
3. A class graph can be made on chart paper or drawn on the chalkboard. A transparency of the student worksheet could also be used for collecting class data.
4. Each group or individual will need only 50 grams of weight.

Procedure
1. Have students make catapults using procedures similar to *May The Force Be With You*. In this activity, they should be experimenting with different fulcrum points in order to attain maximum results. Fifty (50) grams of weight will be held constant for the effort.
2. Instruct students to select their most successful designs and name their catapult entries.
3. Conduct the competition.
4. Gather and record data of class results.
5. Have students complete the graph.

Discussion
1. Was there any relationship in the length of the resistance arms or effort arms and the rankings? Explain.
2. What were some of the reasons that certain catapults worked better than others?

Extensions

What is the lightest weight that must be dropped on the effort side in order for the resistance to be forced off the dowel? From what height must it be dropped? Could we use the mgh formula that we used in the previous lessons to help us discover this?

Major Conceptual Components
- Energy comes in many forms.
- Mechanical energy is the energy due to the position or movement of something.
- Energy cannot be created or destroyed.
- The amount of energy or work put into a machine will come out of the machine in some form.
- Energy is used to accomplish work.
- Work is done when something is moved.
- The amount of work done is determined by how much force was used to move the object and how far the object was moved.
- Levers help to lift loads with less effort. The force needed for lifting changes as the distance of the effort arm changes.

THE GREAT CATAPULT CAPER

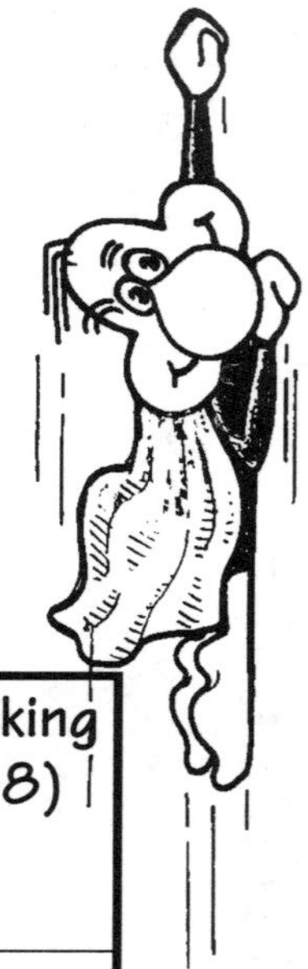

1. Design and name your catapult.
2. Conduct trials, gather data to complete the chart for all entries.
3. Measure the 2 arms of each catapult (distance on either side of the fulcrum).
4. Determine the winners by ranking.

Entry Name	TRIALS			Average	Length of Effort Arm (e)	Length of Resistance Arm (r)	$\frac{r}{e}$	Ranking (1-8)
	1	2	3					
1.								
2.								
3.								
4.								
5.								
6.								
7.								
8.								

5. Was there any relationship between $\frac{r}{e}$ and the rankings? ____________________
6. If so, what was the relationship and why did it occur? ____________________

____

THE GREAT CATAPULT CAPER

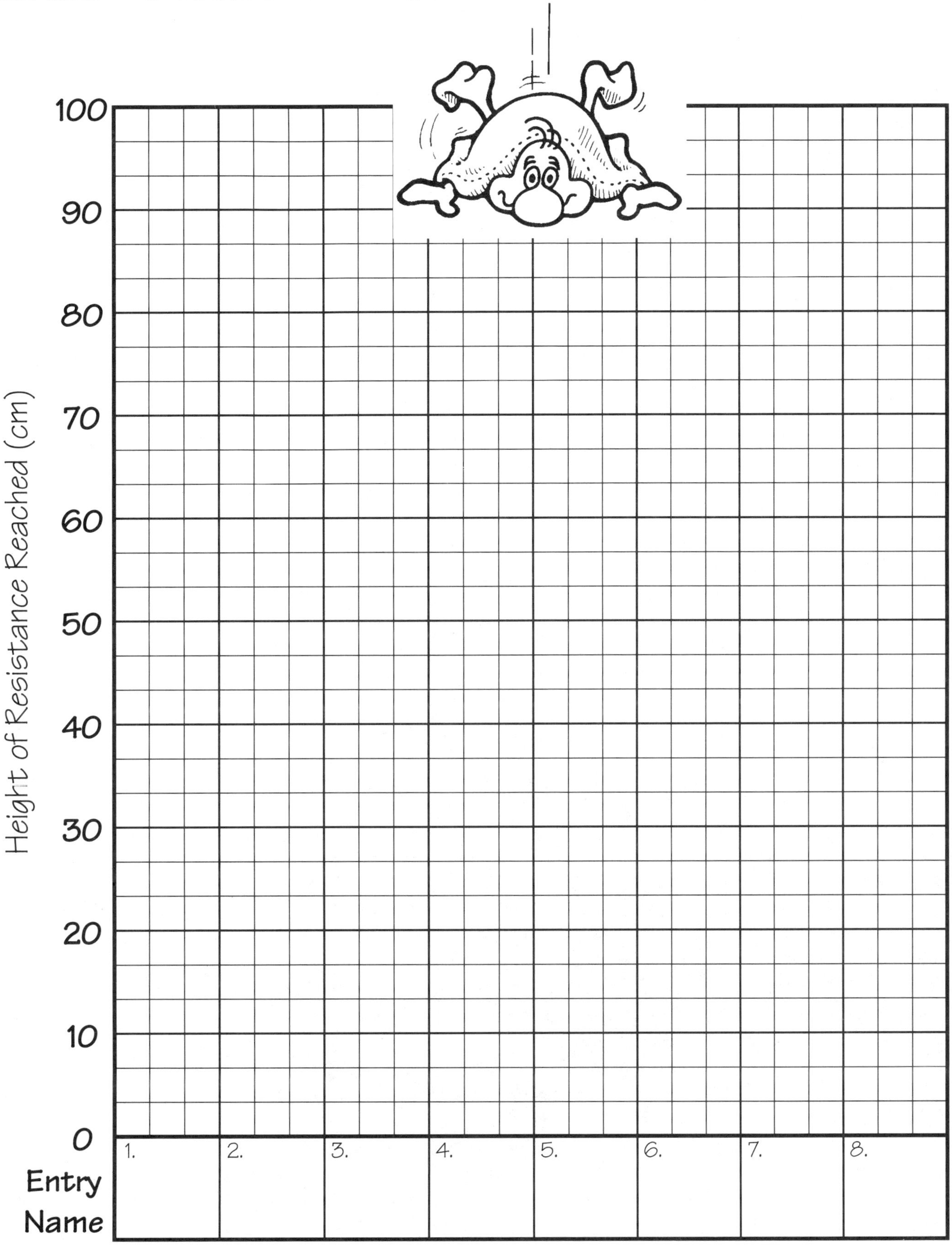

March 7

Dear Journal,

Because of the cold months of winter, it has been awhile since we have been in the shop. Spring seems to be blowing in now. We decided to tune-up our bikes for the coming season. Mucky brought his bike over to the shop and I started cleaning and lubricating everything to discourage Fiendish Friction from getting any strong handholds. When I finished with both our bikes, we took them around the block for a test ride.

Mucky told me he just loves to ride his bike. He said that he just rests his feet on the pedals, they drop, and the bike moves forward. He saw me pedaling and remarked that I had to pedal so fast to go the same speed he did. It was then that I noticed that I was pedaling much faster than Mucky; but, I also noticed that I was moving at the same speed. Hmm, that just can't be. Mucky must have learned a new trick. I decided to get into the shop next week and figure out how this could happen. I wonder if it has to do with the gears. I think this will be the first area I check into.

March 14

Dear Journal,

Because of the cold rains, I stayed inside today and pulled out the box of gears to start exploring. As I began to mesh gears together and turn them, I made a discovery: If a small gear and a large gear were meshed together and I turned one of them, the small gear would always spin more times than the big one. As I kept playing around I also realized that if I turned the big gear once, the small gear would spin around more than once. One the other hand, I found that I had to turn the little gear more than once to get the bigger gear to turn once. I also noticed it was quite a bit harder to turn the big gear than the little gear. I knew from math class that if I were going to figure out this pattern, I was going to have to get organized, so I made a list of my discoveries. Since I was comparing two things, I decided to write out the comparisons as ratios:

Comparison	Size	Speed	Force
Gear that got turned / Gear I turned	smaller / larger	faster / slower	smaller / larger
Gear that got turned / Gear I turned	larger / smaller	slower / faster	larger / smaller

When I turned the big gear, the little gear turned faster but with less force. When I turned the little gear, the big gear turned more slowly and with more force.

I decided to check what the encyclopedia had to say about gears. This is what I found:

> A gear train is a combination of two or more gears used to transmit motion between two rotating shafts. For an ordinary gear train, the ratio of the velocity of the driven gear to the driving gear is called the train value.
>
> $$\text{Train Value} = \frac{\text{velocity of driven gear}}{\text{velocity of driving gear}}$$
>
> For any two gears in mesh, the speeds vary inversely to the number of teeth. Thus, the general expression for the magnitude of the train value can be calculated from the number of teeth.

$$\text{Train Value} = \frac{\text{number of teeth on driving gear}}{\text{number of teeth on driven gear}}$$

Likewise, the ideal torques on the gears varies in direct relationship to the number of teeth and inversely to the speed. Another form of the train value may be written:

$$\text{Train Value} = \frac{\text{torque on driving gear}}{\text{torque on driven gear}}$$

As I studied the encyclopedia and compared my discoveries, I realized they said the same things. You just need to translate those fancy words like velocity and inversely.

This was all making sense to me, but I still wasn't sure how it all applied to Mucky's secret. As I was about to leave, Handy came out to work on some of his tools. While talking to him, he turned to a well-worn page in the encyclopedia and highlighted a couple of sentences I was already familiar with.

Any machine that multiplies force does so at the expense of distance. Any machine that multiplies distance does so at the expense of force.

Handy explained that on Mucky's bike, the bigger gear is on the pedal where Mucky puts the force in. The smaller gear is attached to the rear wheel where the force comes out. Mucky's big, heavy foot pushes slowly on the pedal. His wheel-gear turns faster than his pedal-gear but with a smaller force that is just able to move his bike.

My bike's pedal-gear is smaller in relationship to my wheel-gear. As a result, my little feet have to turn very fast. Even with my faster pedaling, my wheel-gear turns just fast enough to keep up with Mucky, but my gears increase my force more than Mucky's gears.

March 21

Dear Journal,

In celebration of the starting of spring, Mucky and I decided to take a jaunt around town. We decided to make a run up to Look-out Point. Our newly lubricated bikes and pumped-up tires kept us away from the grimy hands of Fiendish Friction. We were feeling pretty good about our progress until we started up the long grade. Then we heard the laugh of Greedy Gravity as she reached out to pull us down. Mucky panicked in fear and pedaled as fast as possible with me chasing after him. Not far up the grade, I caught and passed him only to hear his groans and moans as Gravity slowly wrapped her arms around him, and then dragged him back down the hill, into the ditch at the side of the road. He was still in her clutches when I got to him.

I rolled Mucky to his feet and we walked our bikes home as his was in need of some major repairs. He asked me what my secret was for keeping out of the grasp of Greedy Gravity. I tried to explain that there was no secret, but that it was all in the bike's gears. I told him that his gears gave him more speed so he could pedal slower and go just as fast as me, but that with my gearing I had to pedal faster to stay with him which meant I got more force so I could pull away from Greedy Gravity.

Mucky couldn't make sense of what I was saying. I'll just have to get out the gears in the shop and show him.

One Good Turn Deserves Another

Topic Area
Wheels and belts

Introductory Statement
Students will discover what happens when two or more wheels are joined by a belt or belts.

Math
Using rational numbers
ratios
proportions
Using positive and negative integers
Estimating

Science
Physical science
Simple machines

Math/Science Process Skills
Predicting
Observing
Classifying
Inferring
Collecting and recording data
Identifying and controlling variables
Interpreting data
Applying

Materials
Per student or group:
four thread spools, all with same diameter
1/4" 10 cm X 25 cm pegboard pieces
three rubber bands–size is relative to spool size (For large spools use a rubber band that is approximately 9 cm X 0.5 cm)
four 2" pieces of 1/4" doweling
optional: paraffin wax or soap bars

Key Questions
1. How are the turns of wheels affected by the configuration of a belt?
2. What happens to the wheels when you connect multiple wheels with belts in differing configurations?

Background Information
When two wheels, or pulleys, are connected with a belt, the belt makes it possible for one wheel to turn another. If the belt is placed so an oval is formed, both wheels will move in the same direction. If one wheel is turned clockwise (CW), the second wheel will turn clockwise (CW). Or, vice versa, when one wheel is turned counterclockwise (CCW), the second wheel will turn counterclockwise (CCW).

If the belt is placed to form a figure eight, the direction the wheels travel in relation to each other will be opposite to each other. In this case, when one wheel is turned CW, the other wheel will turn CCW. Or, when one wheel is turned CCW, the second wheel will turn CW.

In this investigation, the wheel that is manually turned is identified as the *power wheel*.

Management
1. Several days prior to this investigation, ask the students to bring from home four thread spools. Although the entire class does not have to use spools of the same diameter, each student or group will need to have four spools of the same size.
2. The standard hole size on the spools is a little larger than 1/4". This will allow the spool to spin freely on the 1/4" dowel.
3. Cut the dowel into two inch lengths.
4. A piece of pegboard should be provided for each student. If the class is working in groups–and groups of two work best–one piece of pegboard is needed for each group.
5. Prior to this investigation, students should have an understanding of the properties of friction, its causes and prevention by the use of lubricants. It is suggested that students first do the activity *Start Your Engines*.
6. During the investigation, students may discover a certain amount of belt slippage. The slippage may be a result of a number of variables: friction; the thickness of the rubber band; or the condition, age, and tension of the rubber band. (These factors can stimulate a good discussion of the importance of regularly checking the belts on cars and trucks.) Students may want to use paraffin wax or soap bars as a lubricant to reduce the friction.

Procedure
1. Distribute materials to each group.
2. Ask *Key Questions*.
3. Have the students place the four dowel pieces in the positions on the peg board as shown in the illustration.
4. Instruct the students to mark their spools by drawing a straight line from the center of each spool to its outer edge. These lines will serve as markers to aid the counting of the spools' rotations.
5. Inform the students that while doing the worksheets, the power wheel, the wheel turned by hand, will be spool A.
6. Using the worksheet *One Good Turn Deserves Another*, have the students solve the first four problems and then complete the ratio chart at the end. Since the wheels are the same diameter, the ratio will be 1:1. However, there may be a certain amount of slippage of the belts that could distort the ideal data.
7. Have students complete the rest of the worksheets keeping in mind and allowing for the slippage of the belts.

Discussion

1. How do the effects of oval connections differ from figure eight connections? [with oval connections, the wheels rotate the same direction; with figure eight connections, the wheels rotate in opposite directions]
2. What general rule can you make about the arrangement of wheels and belts? [If the belt makes an oval connection and wheel A turns CW, wheel B will turn CW. If the belt makes a figure eight connection and wheel A turns CW, wheel B will turn CCW.]
3. Make a symbol system to determine what direction the last wheel turns in relation to the first. [Answers will vary. One solution: let + indicate the same direction and - indicate opposite directions. Give each connection its symbol equivalent. The sum of the symbols is the final outcome.]
4. How many belt patterns could be used to turn A clockwise and D counterclockwise? What are those patterns?
5. What generalization can you make from the turn chart? [this turn ratio chart indicates that one turn in gives one turn out]

Extensions

1. Have students go on a visual scavenger hunt to find examples of wheels (pulleys) and belts being used in machinery.
2. Ask students to construct a demonstration of wheels working to change direction.
3. Study the Mobius strip. Power takeoff belts are often connected as a Mobius strip. Determine how the Mobius strip would affect the direction of the wheels. What is the advantage of this type of belt?

Major Conceptual Components

- Energy comes in many forms.
- Mechanical energy is the energy due to the position or the movement of something.
- A machine is a device for multiplying forces or changing the direction of forces.
- Wheels and axles are turning levers.

ONE GOOD TURN DESERVES ANOTHER

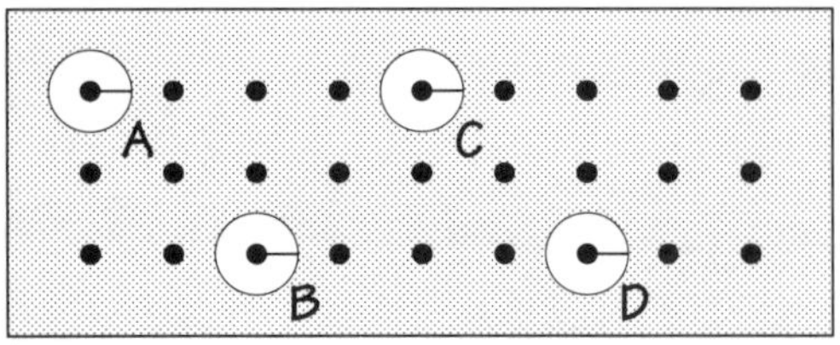

Arrange your board as shown above. Then connect A to B with a rubber band as shown in each picture. Turn A the way shown by the arrow and record which direction B turns.

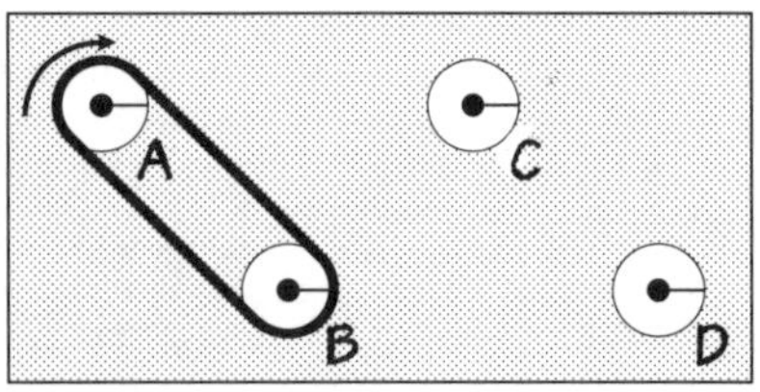

If A turns clockwise (CW)

B turns ______________

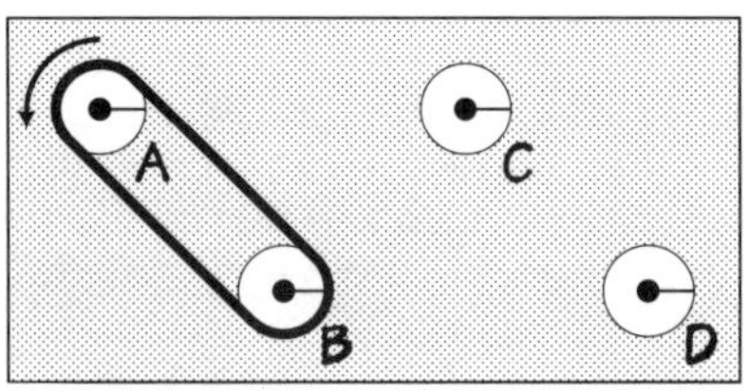

If A turns counterclockwise (CCW)

B turns ______________

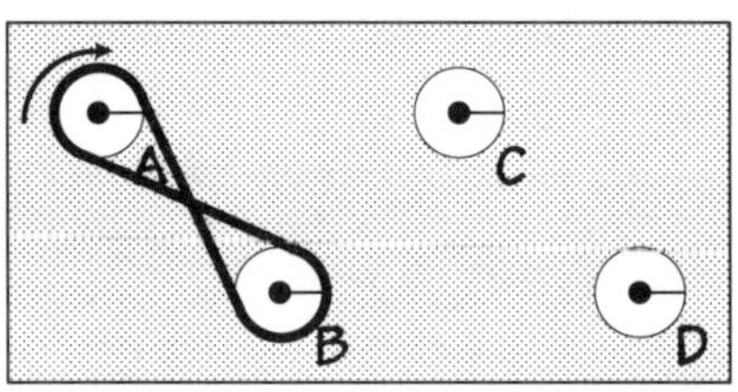

If A turns clockwise (CW)

B turns ______________

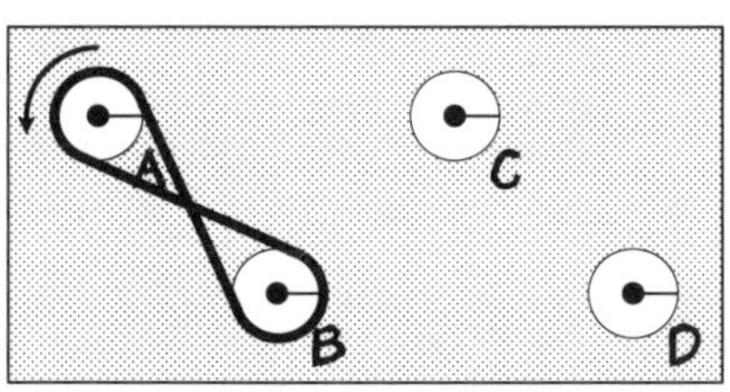

If A turns counterclockwise (CCW)

B turns ______________

Turn A the number shown on the chart and record what happens to B.

Turn Chart

Turns of A	1	2	3
Turns of B			

ONE GOOD TURN DESERVES ANOTHER

Connect A to B and B to C using 2 rubber bands as shown in each picture.
Turn A the way shown by the arrow and record which direction B and C turn.

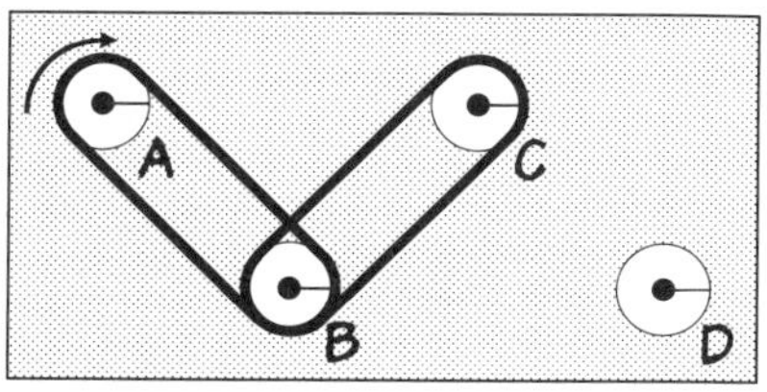

If A turns CW
B turns ______
C turns ______
If C turns CCW
A turns ______

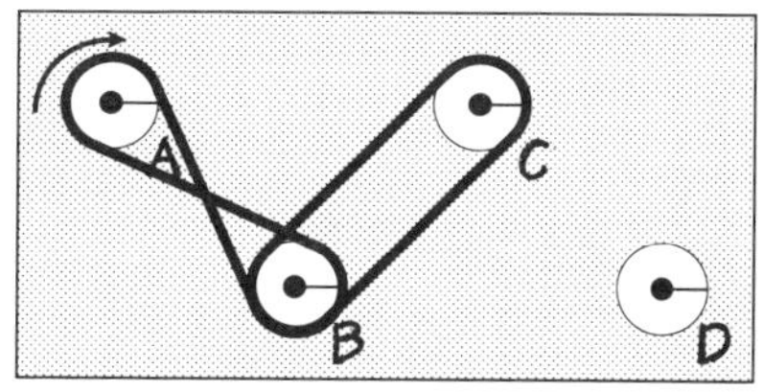

If A turns CW
B turns ______
C turns ______
If C turns CCW
A turns ______

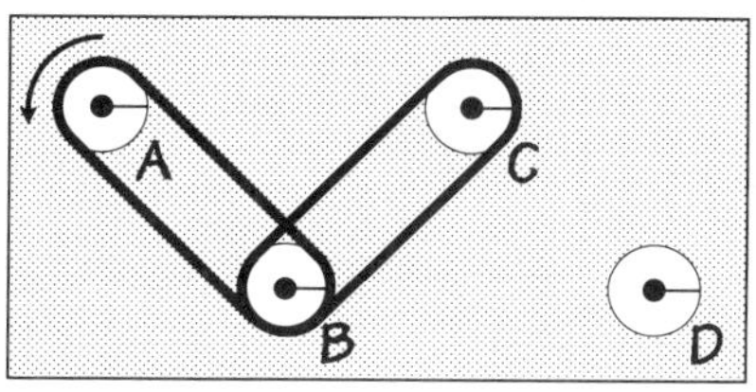

If A turns CCW
B turns ______
C turns ______
If C turns CW
A turns ______

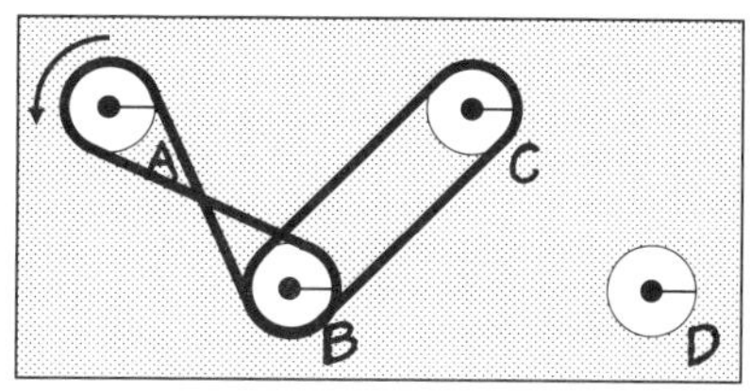

If A turns CCW
B turns ______
C turns ______
If C turns CW
A turns ______

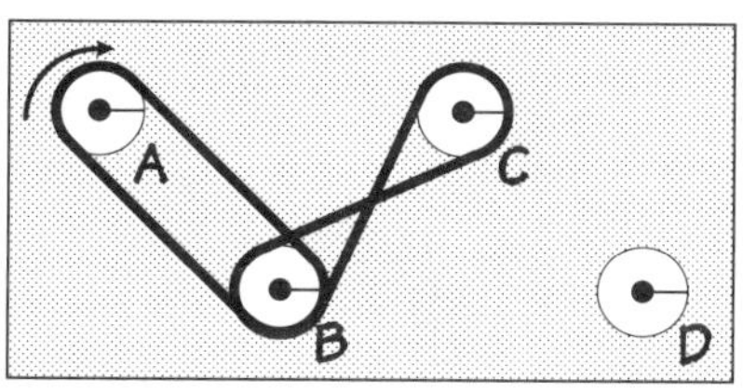

If A turns CW
B turns ______
C turns ______
If C turns CW
A turns ______

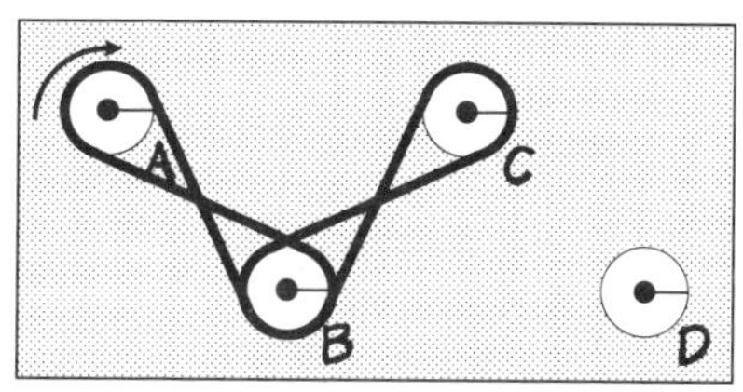

If A turns CW
B turns ______
C turns ______
If C turns CW
A turns ______

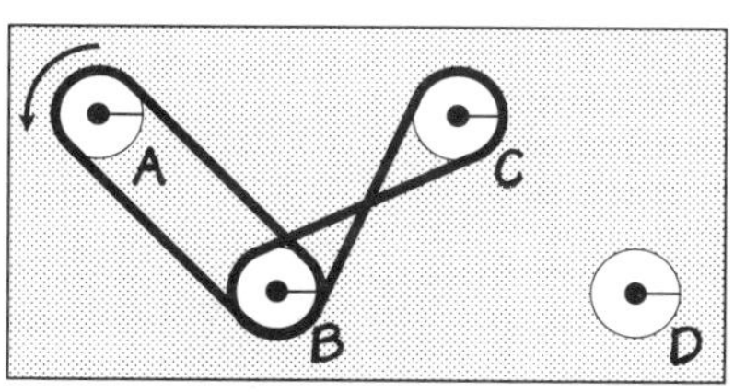

If A turns CCW
B turns ______
C turns ______
If C turns CCW
A turns ______

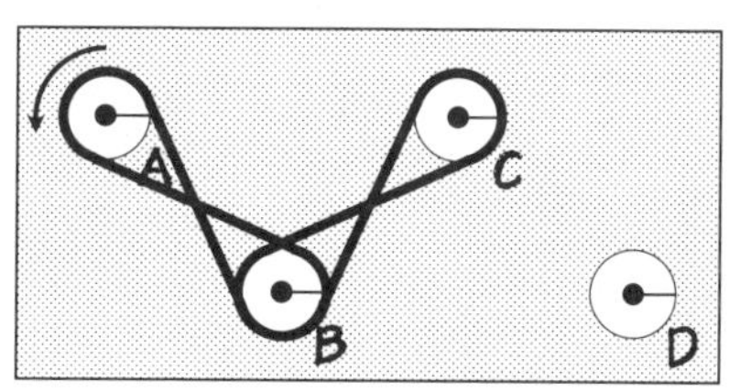

If A turns CCW
B turns ______
C turns ______
If C turns CCW
A turns ______

Turn A the number shown on the chart and record how many turns B and C make.

Turn Chart

Turns of A	1	2	3
Turns of B			
Turns of C			

ONE GOOD TURN DESERVES ANOTHER

Use 3 rubber bands and construct each system. Turn A and record what happens.

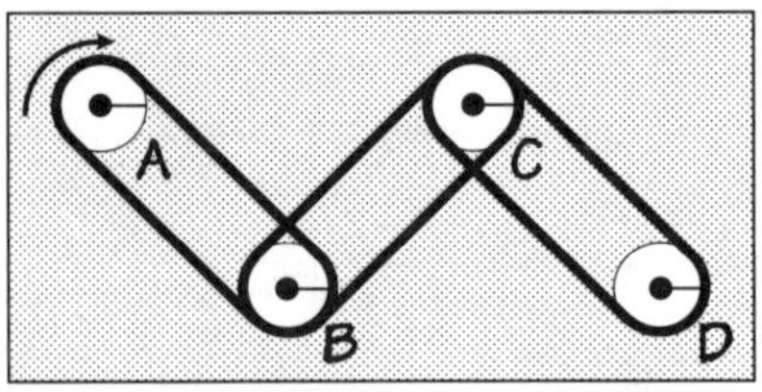

If A turns CW
C turns ______
D turns ______

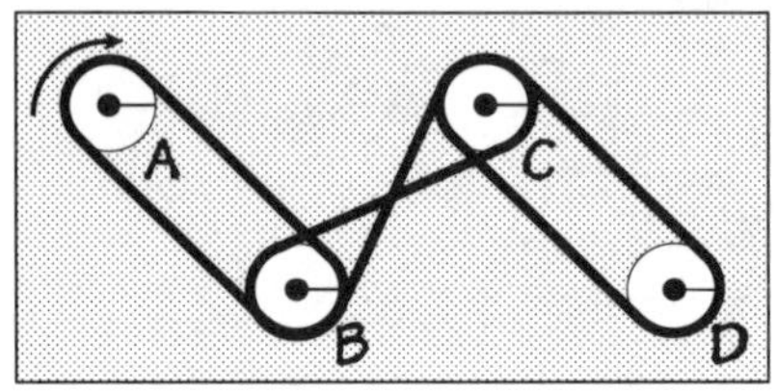

If A turns CW
C turns ______
D turns ______

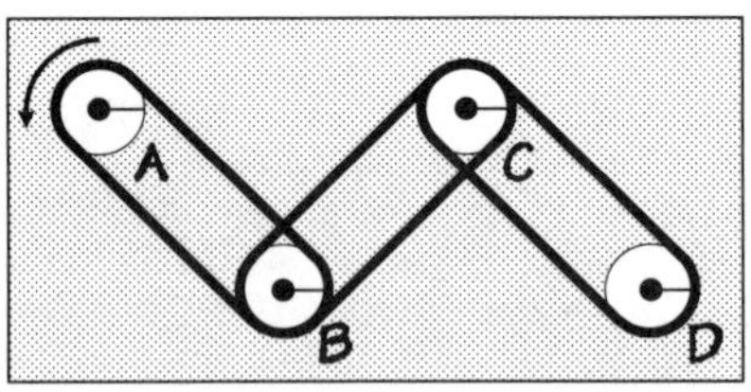

If A turns CCW
C turns ______
D turns ______

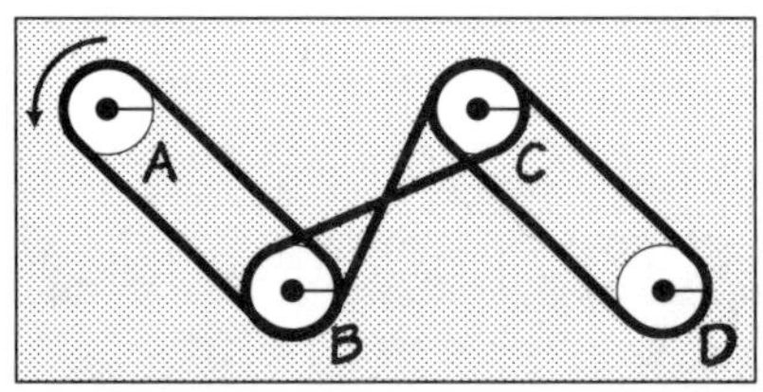

If A turns CCW
C turns ______
D turns ______

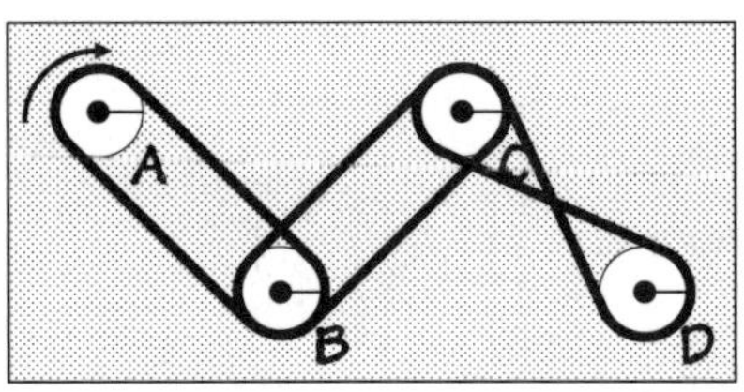

If A turns CW
C turns ______
D turns ______

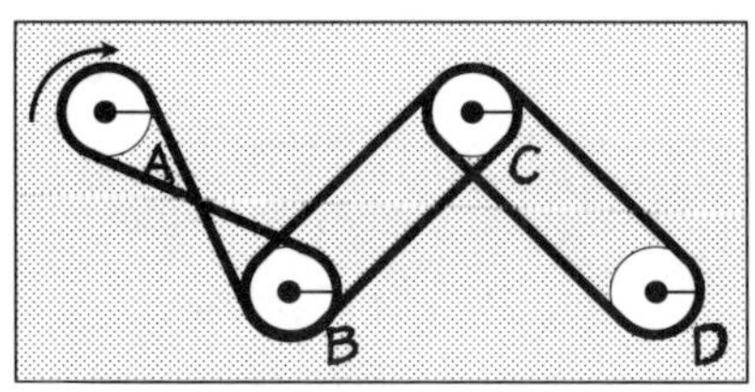

If A turns CW
C turns ______
D turns ______

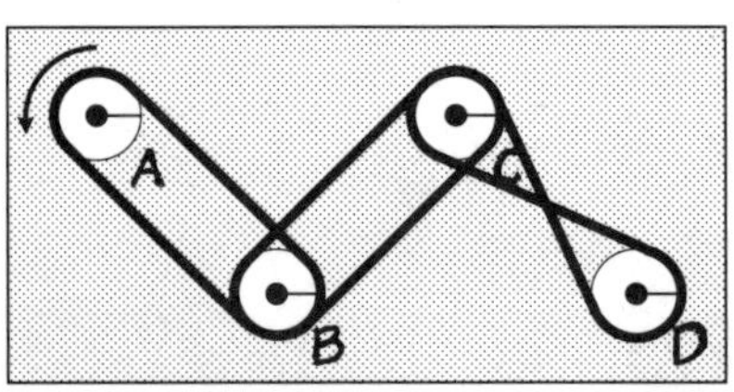

If A turns CCW
C turns ______
D turns ______

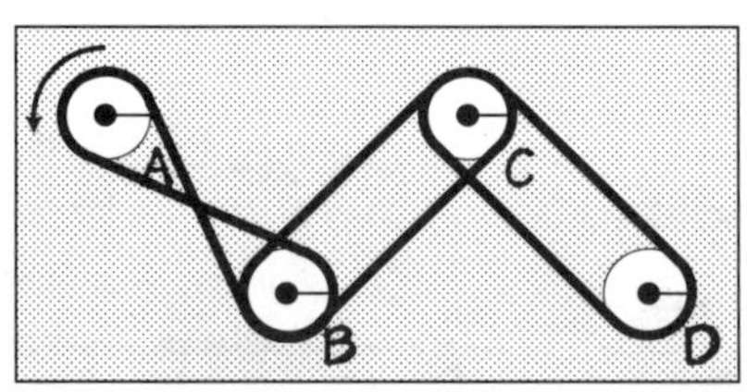

If A turns CCW
C turns ______
D turns ______

ONE GOOD TURN DESERVES ANOTHER

Use 3 rubber bands and construct each system. Turn A and record what happens.

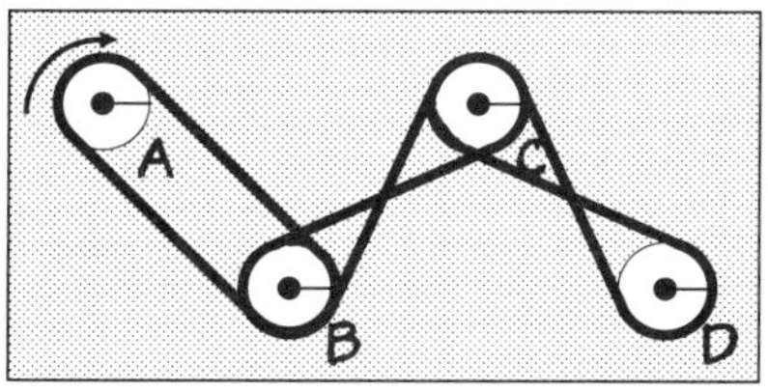

If A turns CW
B turns ______
C turns ______
D turns ______

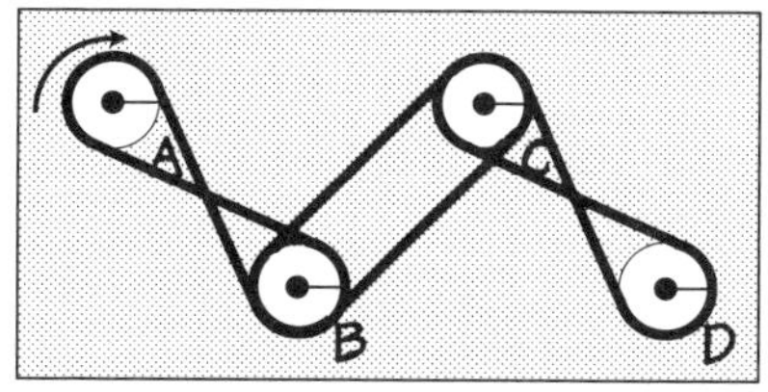

If A turns CW
C turns ______
D turns ______

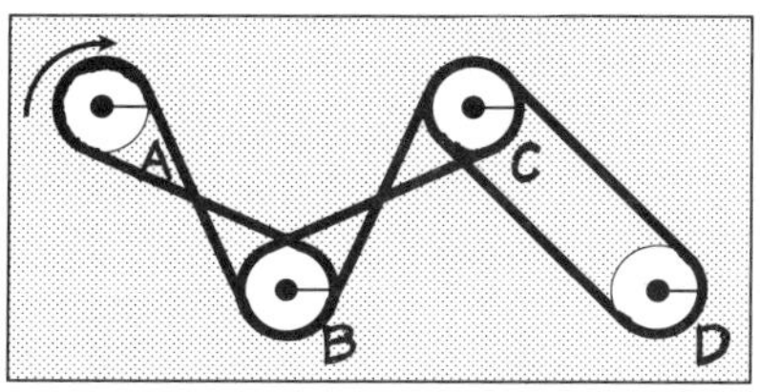

If A turns CW
B turns ______
C turns ______
D turns ______

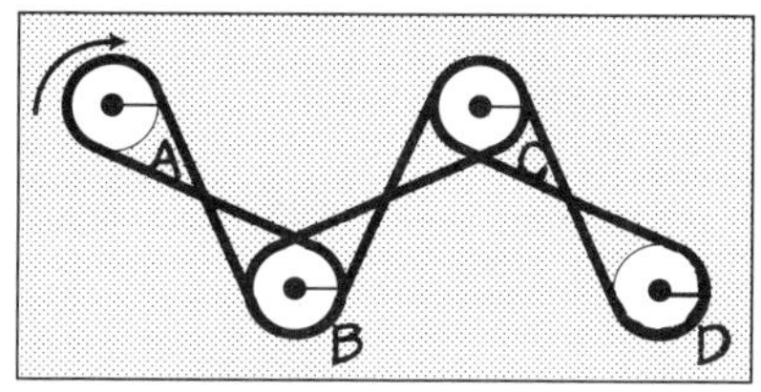

If A turns CW
C turns ______
D turns ______

Turn Chart

Turns of A	1	2	3
Turns of B			
Turns of C			
Turns of D			

Turn A the number shown on the chart and record how many turns B, C, and D make.

How many belt patterns could be used to turn A clockwise and D clockwise?
What are the patterns?

What would be the effect on the turn chart if the wheels were of different diameters?

Gearing Up

Topic Area
Gears and tooth ratios

Introductory Statement
Students will construct gears. Using these gears, they will discover the rotational pattern of meshed gears and gain a basic understanding of the inverse relationship of teeth and turn ratios.

Math
Using logical reasoning
Analyzing math patterns
  inverse relationships
Using rational numbers
  fractions
  ratios

Science
Physical science
  Simple machines

Math/Science Process Skills
Making and testing hypotheses
Predicting
Observing
Classifying
Comparing and contrasting
Applying and generalizing

Materials
For each group:
  8 push pins
  cover stock or tagboard
  9 X 11 inch corrugated cardboard
  scissors
  glue

For the class:
  optional: egg beater

Key Question
When you turn one gear in a pair, how can you predict the action of the other gear?

Teacher Background
Gears are wheels with teeth which make it possible for one wheel to turn another without the use of a belt. Gears are used for the same purposes as belt and wheel systems (pulleys). They are used when the slippage of a belt system needs to be avoided as in the precision of a watch or in the high power transfer of a car. A gear is a special form of a wheel and axle.

Gears are used for four main purposes: to transfer power, to change direction of rotations, to change the speed of rotation, and to increase the rotational force (torque). In this activity, students will explore the first three purposes.

Gears which are linked together with chains will have a rotational pattern the same as belt and wheel systems. If gears are linked by meshing, their rotational pattern will change–the gears will rotate in opposite directions. If a series of gears are laid out and meshed in a sequence, the first, third, and any other odd-sequenced gears will rotate in the same direction (clockwise or counterclockwise). The second, fourth, and all even-sequenced gears will rotate in the same direction, but that direction will be opposite the odd-sequenced gears.

The relationship between the number of teeth on two gears determines the change in speed and force of rotations. In this activity, students will be using gears with 10 teeth and 20 teeth. The 10-tooth gear and the 20-tooth gear will mesh with one another in a gear train. In this example, consider the smaller gear as the *drive gear*–the gear that the student will turn. If the drive gear is rotated exactly once, 10 teeth on the small drive gear will mesh with 10 teeth on the larger *driven gear*. The smaller gear has only 10 teeth so it must rotate twice to mesh with the 20 teeth on the larger gear. The driven gear makes only half a rotation. Inversely, since the driven gear rotates half as much, it has twice the force with each rotation. The driven gear has twice the teeth, rotates half as much, and has two times the force of the drive gear.

By knowing the number of teeth on a pair of meshing gears, one can determine the change in the speed (amount of rotation) and the mechanical advantage of a set of gears. In the example, the teeth ratio is 20/10, or 2/1; the turn ratio is the inverse of the teeth ratio, or 1/2; and the mechanical advantage (a ratio of output/input) is the same as the teeth ratio, or 2–both are the inverse of the turn ratio.

Management
1. Groups of three or four work well for this activity. Each student can be responsible for several gears.
2. If you have access to a copy machine that will print on cover stock, copy the gears on cover stock before class. You may want to make double copies as the gears may work better by using a double thickness.
3. The ease at which the gears turn each other depends on the care taken when making the gears.
4. Students will need to help both gears turn to insure that the teeth do not skip.

Procedure
1. Discuss the *Key Question.*
2. Have the students glue the two gear pattern sheets onto pieces of tagboard if cover stock copies are not available.
3. Have students carefully and accurately cut out the gears.
4. The students should fold up the gear teeth along the dotted line so that they are raised about 0.5 cm above the gear face.
5. Have students cut out a 9 X 11 inch piece of cardboard to serve as a base for attaching the gears.

6. Attach gear G to the cardboard base using a push pin through its center.
7. Insert a pin through the center of gear A and place this gear on the base so that gears G and A mesh. The students may have to move gear A from side to side until the teeth of the gears come together. Pin the gear to the base.
8. Distribute *Teeth and Turns* handout.
9. Have students match the black teeth on the gears. They then turn gear G clockwise until both black teeth come back to the starting positions. The students will count the number of rotations required by each gear to return to this starting position and record the results on the worksheet. One student should count the turns made by the drive gear while another counts the turns made by the driven gear. Warn students to turn the gears carefully so as not to damage them.
10. Continue this process for all eight gear trains listed on the handout.
11. Ask students to look for patterns in the teeth ratios and turn ratios. Discuss these. If students do not discover the inverse relationship, distribute the *Fraction Families* handout. Have students complete the handout to help them discover the inverse relationship of teeth and turn ratios.
12. Distribute *All Aboard for the Next Train* and *Driven to Distraction* handout.
13. Have students complete *All Aboard for the Next Train* by listing all the possible gear teeth combinations. The combinations on this handout are then used for selecting gear combinations on the *Driven to Distraction* handout in which the students predict the turn ratio from the teeth ratio. They will then test the turn ratios to check their predictions.
14. Distribute *The Great Turn Around.* Instruct the students to complete this handout by making a long sequence of meshed gears and predicting which way each gear will turn if the first gear is turned clockwise. Test the predictions and record the actual results.

Discussion

1. With a gear train, how do you determine how many turns each gear will make to return both gears to their starting positions? [Use the inverse of the teeth ratio in simplest form]
2. What are the fastest and slowest turn ratios possible with these gears? [fastest: H/A which has a ratio of 5/2 or 2.5; slowest: A/H which has a ratio of 2/5 or 0.4]
3. How do different sized gears change the speed of a bicycle? [different teeth ratios give different turn ratios]
4. An egg beater has gears that increase the speed and also change the direction. Can you use an egg beater to demonstrate this?
5. What are some other things around your house that use gears? How do the gears change the speed or direction of rotations?
6. In a clock, all three hands (the second, the minute, and the hour hand) rotate in the same direction at different speeds, how might their gears be connected?

Extensions

Have students test to see if placing more than two gears in sequence changes the speed more than using just the two gears. [Only the drive and final driven gear matter. The gears in-between will turn at their respective ratios to the drive gear. The gears between the drive and final driven gear are called idlers.]

Curriculum Correlations

SUNBURST, INC®. produces a computer program called *Gears* that will reinforce and may be used in conjunction with this activity.

Major Conceptual Components

- Energy comes in many forms.
- Mechanical energy is the energy due to the position or the movement of something.
- A machine is a device for multiplying forces or changing the direction of forces.
- Wheels and axles are turning levers.
- Gears are toothed wheels.
- Gears can be used to transfer forces from one part of a machine to another.

Gearing Up...Gears

Gearing Up...Gears

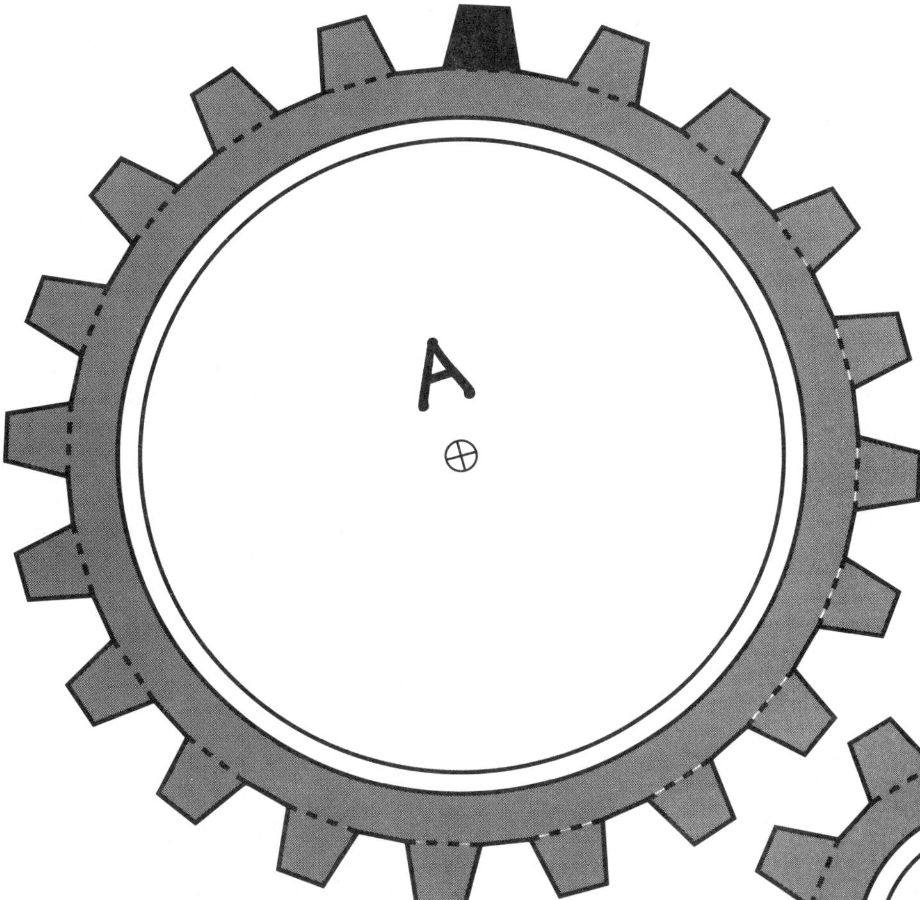

1. Paste the gear pattern sheet on to a piece of tagboard.
2. Carefully and accurately cut out the gears along the dark line.
3. Fold up the gear teeth along the dotted line so that they are raised about 0.5 cm above the gear face.
4. Cut out 9"x11" cardboard to serve as a base for attaching the gears.
5. Push a pin through the center of a gear and then into the cardboard base.

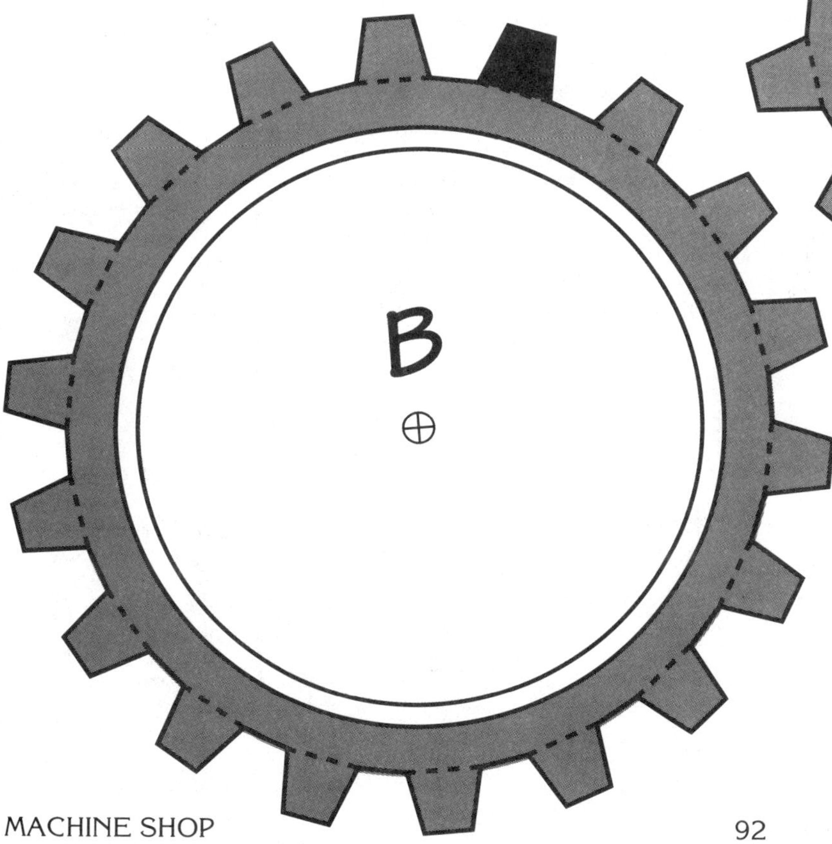

6. Insert a pin through the center of another gear and place this gear on the base, moving it until the teeth of the two gears come together (mesh). Match the black teeth of the two gears. Pin the gear to the base.
7. Turn the drive gear clockwise until both black teeth are back in their starting positions. Count how many times each gear turns.

TEETH AND TURNS

1. Pin the drive gear onto the cardboard base.
2. Pin the output gear to the base, moving it from side to side until the teeth of the two gears come together (mesh).
3. Match the black teeth on the gears. Turn the drive gear clockwise until both black teeth are back in their starting positions. Record how many times each gear turned.

Train 1	Teeth Ratio	Turn Ratio
Driven: A	___	___
Drive: G		

Train 2	Teeth Ratio	Turn Ratio
Driven: C	___	___
Drive: H		

Train 3	Teeth Ratio	Turn Ratio
Driven: F	___	___
Drive: H		

Train 4	Teeth Ratio	Turn Ratio
Driven: B	___	___
Drive: F		

Train 5	Teeth Ratio	Turn Ratio
Driven: D	___	___
Drive: G		

Train 6	Teeth Ratio	Turn Ratio
Driven: D	___	___
Drive: F		

Train 7	Teeth Ratio	Turn Ratio
Driven: G	___	___
Drive: H		

Train 8	Teeth Ratio	Turn Ratio
Driven: A	___	___
Drive: H		

What relationships do you notice between the turn ratio and teeth ratio for each gear set?

FRACTION FAMILIES

1. Complete the lists of equivalent fractions below.

Fraction Family of 1/2 : $\frac{1}{2} = \frac{2}{4} = \frac{3}{6} = \frac{4}{8} = \_ = \_ = \_ = \_ = \_ = \_ = \frac{11}{22}$

Fraction Family of 2/3 : $\frac{2}{3} = \frac{4}{6} = \frac{6}{9} = \_ = \_ = \_ = \_ = \_ = \_ = \_ = \frac{22}{33}$

Fraction Family of 4/5 : $\frac{4}{5} = \frac{8}{10} = \_ = \_ = \_ = \_ = \_ = \_ = \_ = \_ = \frac{44}{55}$

Fraction Family of 2/5 : $\frac{2}{5} = \frac{4}{10} = \_ = \_ = \_ = \_ = \_ = \_ = \_ = \_ = \frac{22}{55}$

2. Circle any fractions above that are reciprocals of any teeth ratios on the record sheet.

3. Record every circled fraction by the gear train of its reciprocal on the *Teeth and Turns* handout. Write these fractions in the pictures of the gears by the teeth and turns ratios.

4. What relationship do you notice between the turn ratio and the fraction family of the teeth ratio's reciprocal?

5. How could you use this relationship to determine the turn ratio of two gears without turning them?

All Aboard for the Next Train

Make a list of the number of teeth on gears A-H.

_____, _____, _____, _____, _____, _____, _____, _____.

Using these numbers, list the teeth ratios for all the possible gear trains.

$$\frac{\text{Driven}}{\text{Drive}} = \frac{10}{8}, \frac{12}{8}, \frac{14}{8}, \frac{\_\_}{8}, \frac{\_\_}{8}, \frac{\_\_}{8}, \frac{\_\_}{8}$$

$$\frac{\text{Driven}}{\text{Drive}} = \frac{8}{10}, \frac{12}{10}, \frac{\_\_}{10}, \frac{\_\_}{10}, \_\_, \_\_, \_\_$$

$$\frac{\text{Driven}}{\text{Drive}} = \frac{8}{12}, \frac{10}{12}, \_\_, \_\_, \_\_, \_\_, \_\_$$

$$\frac{\text{Driven}}{\text{Drive}} = \frac{\_\_}{14}, \_\_, \_\_, \_\_, \_\_, \_\_, \_\_$$

$$\frac{\text{Driven}}{\text{Drive}} = \frac{\_\_}{15}, \_\_, \_\_, \_\_, \_\_, \_\_, \_\_$$

$$\frac{\text{Driven}}{\text{Drive}} = \frac{\_\_}{16}, \_\_, \_\_, \_\_, \_\_, \_\_, \_\_$$

$$\frac{\text{Driven}}{\text{Drive}} = \frac{\_\_}{18}, \_\_, \_\_, \_\_, \_\_, \_\_, \_\_$$

$$\frac{\text{Driven}}{\text{Drive}} = \frac{\_\_}{20}, \_\_, \_\_, \_\_, \_\_, \_\_, \_\_$$

Driven to Distraction

1. Choose 6 gear trains from the list of all possible trains which you have not tried.

2. Complete the teeth ratio and predict how many turns will be needed to return the gears to their starting positions.

3. Mesh the two gears of the train and then turn the drive gear to test how many turns are actually needed.

Train 9	Teeth Ratio	Predicted Turn Ratio	Actual Turn Ratio
Driven:	___	___	___
Drive:			

Train 10	Teeth Ratio	Predicted Turn Ratio	Actual Turn Ratio
Driven:	___	___	___
Drive:			

Train 11	Teeth Ratio	Predicted Turn Ratio	Actual Turn Ratio
Driven:	___	___	___
Drive:			

Train 12	Teeth Ratio	Predicted Turn Ratio	Actual Turn Ratio
Driven:	___	___	___
Drive:			

Train 13	Teeth Ratio	Predicted Turn Ratio	Actual Turn Ratio
Driven:	___	___	___
Drive:			

Train 14	Teeth Ratio	Predicted Turn Ratio	Actual Turn Ratio
Driven:	___	___	___
Drive:			

If all you know is the number of teeth on the gears, explain how you determine how many turns each gear will make to return both gears to the starting position.

The Great Turn Around

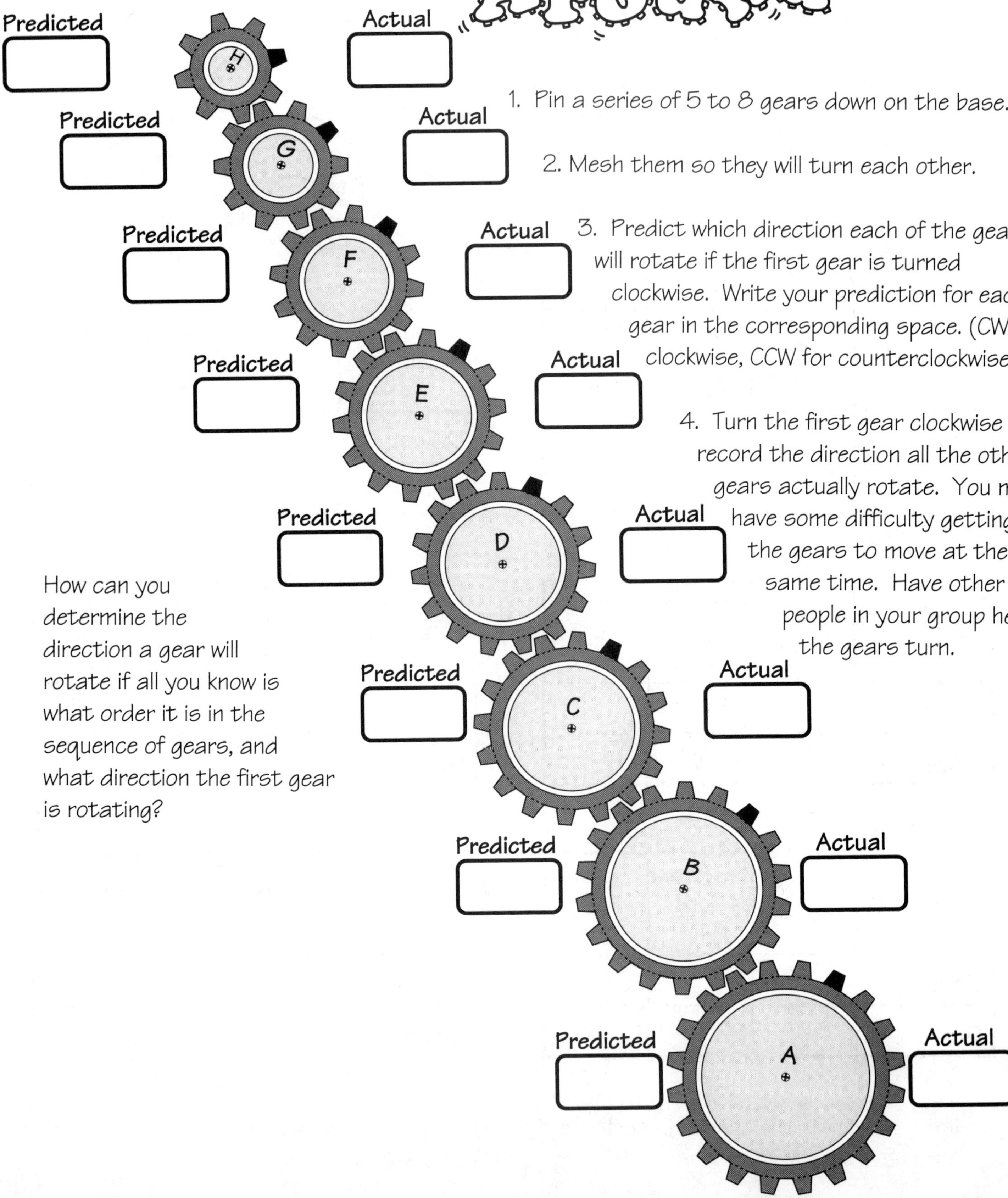

1. Pin a series of 5 to 8 gears down on the base.

2. Mesh them so they will turn each other.

3. Predict which direction each of the gears will rotate if the first gear is turned clockwise. Write your prediction for each gear in the corresponding space. (CW for clockwise, CCW for counterclockwise)

4. Turn the first gear clockwise and record the direction all the other gears actually rotate. You may have some difficulty getting all the gears to move at the same time. Have other people in your group help the gears turn.

How can you determine the direction a gear will rotate if all you know is what order it is in the sequence of gears, and what direction the first gear is rotating?

Tooth to Tooth

Topic Area
Wheel and axle, gearing

Introductory Statement
In this activity students will determine the relationship between the gears of a bike and the speed and force of the pedals and wheels. Students will first count the teeth on the gears of a bicycle. They will then turn the bike's pedal and count the turns of the wheel comparing them to the turns of the pedal.

Math
Using geometry
Using rational numbers
ratios
Measuring
angles

Science
Physical science
Simple machines

Math/Science Process Skills
Observing
Collecting and recording data
Applying and generalizing

Materials
Per group:
10-speed bicycle
scrap blocks of wood
scissors
tape

Key Question
How can you determine how much a bike's wheel will turn by the size of the gears?

Background Information
A ten-speed bike has two gears at the pedal, and five gears on the axle of the rear wheel. This provides ten possible combinations: five with the large pedal gear, and five with the smaller pedal gear. Both the pedal gears are larger than any of the wheel gears. Because the pedal gears are larger, each rotation generates more then one rotation of the wheel gear. By comparing the number of teeth on the pedal gear to the number of teeth on the wheel gear, the turning rate can be determined. As an example: A pedal gear has 72 teeth and the wheel gear has 18 teeth. When the pedal rotates once, 72 links of chain are moved. These 72 links in turn move 72 teeth on the wheel gear. To use up 72 teeth, the wheel gear must rotate four times. (72:18 = 72/18 = 4/1 = 4:1) This is an inverse relationship: The pedal is four times as big which makes the wheel gear turns four times as fast.

The fastest gear combination is the largest pedal gear with the smallest wheel gear. The slowest combination is the smallest pedal gear with the largest wheel gear. The fastest combination is also the hardest to pedal. The slowest combination is the easiest to pedal. This is because the gear is a modification of the simple machine, the wheel. As with all simple machines, a trade is made: To get higher speed or longer distance from each pedal rotation, the force is "traded" for a smaller force.

Management
1. This activity works best with small groups of students.
2. Each group of students will need to have access to a 10-speed bicycle for this activity. Students, may be asked to volunteer to bring bicycles into the classroom. If not enough bicycles are available, the teacher may choose to set up a center where a group of students can work.
3. This activity takes two or three, 50-minute class periods to complete: one period during which students discuss the problem and count the teeth on each of the gears; the second and third periods will involve the counting of gear turns while cranking the pedal, and then analyzing the data.
4. The teacher and students must become familiar with the operation of a 10-speed bicycle before beginning this activity. Make sure each group of students has identified which lever operates the front and the rear derailleurs, the gear mechanism that changes the gear ratio, by moving the bicycle chain from one sprocket to another. Students need to be instructed that gears should be changed only when the bicycle is being pedaled.
5. There should be a discussion about safety which includes the dangers of putting hands and objects into the spokes. Students should be encouraged to secure long hair and loose clothing so it does not get caught in the moving parts.

Procedure
1. Divide students into groups. A minimum of four students is required for each group: one person to steady the bike; one to crank the pedal; another to apply friction to the wheel with a board so it does not freewheel; and the fourth member to measure the rotations of the wheel and record the data.
2. Discuss the *Key Question* with students.
3. Have students count the teeth on each gear, record these numbers on the chart, and calculate the teeth ratios. If there are more gears on the bike than allowed for on the chart, have the students count and record the smallest ones.
4. Distribute the two protractor sheets to each student group. Direct the students to cut out the protractor and mount it as instructed onto the rear wheel of the bicycle.

5. Instruct students to turn the bike upside down and place it so it rests upright on the seat and handle bars. One group member will hold it in this position by holding the front fork and handle bars.
6. A second group member turns the rear wheel so that the zero mark on the protractor is in line with a bar of the bicycle frame. That member applies friction to the rear tire with a scrap of wood so that it does not free-wheel when it is pedaled.
7. A third group member rotates the pedal backwards without turning the rear wheel until the crank is in line with one of the bars of the frame. (You should notice a clicking sound, but the wheel will not move.)
8. When the pedal crank and the rear wheel are both in the starting position, the student at the pedal slowly turns the pedal one complete rotation. This turns the wheel to which friction is being applied. The fourth group member measures how many degrees of rotation the rear wheel makes for the pedal's complete rotation. This measurement is recorded on the chart and the decimal equivalent calculated.
9. The gear arrangement is changed and steps 5-8 are followed until all the possible arrangements have been studied.
10. Have students draw conclusions from the activity and discuss their findings.
11. If it is appropriate, students may go outside to ride the bicycles and discover that the "Biggest=Fastest" gear arrangement takes the greatest force to move, while the "Smallest=Slowest" arrangement is the easiest to move. The bicycle gears trade force for speed like all simple machines.

Discussion

1. What does each decimal equivalent tell you about the pedal gear and wheel gear? [How many times bigger the pedal gear is.]
2. What does each decimal equivalent tell you about the turns of the pedal and the rear wheel? [The number of turns the rear wheel makes for each pedal turn.]
3. What similarities do you find in the two decimal equivalents for each gear arrangement? [If the activity were done accurately, they will be nearly the same.]
4. How could you figure out the number of turns the back wheel would make if you only knew the number of teeth on the pedal and wheel gears? [Use the ratio of teeth on the pedal gear to teeth on the wheel gear to determine the inverse relationship.]

Extension

Have students determine the circumference of the rear tire and make a chart that shows how far the tire travels for each rotation of the pedal by multiplying the decimal equivalent by the circumference. They could go outside and see how accurately this works for measurement by "pedaling" off distances they have measured. Students will need to realize this works only if they do NOT let the bike glide and free wheel.

Major Conceptual Components

- Energy comes in many forms.
- Mechanical energy is the energy due to the position or the movement of something.
- A machine is a device for multiplying force or changing the direction of force.
- Wheels and axles are turning levers.
- Gears are toothed wheels.
- Gears can be used to transfer force from one part of a machine to another.

TOOTH to TOOTH

How can you determine how much a bike's wheel will turn by the size of the gears?

Count the number of teeth on each gear and record it on the chart. Then determine the teeth ratio and decimal equivalent for each combination.

		Largest Pedal Gear = A Number of Teeth = ______			Smallest Pedal Gear = B Number of Teeth = ______		
		Gears Connected	Teeth Ratio	Decimal Equivalent	Gears Connected	Teeth Ratio	Decimal Equivalent
Wheel Gears	Smallest gear = 1 Teeth = _____	A / 1			B / 1		
	Gear 2 Teeth = _____	A / 2			B / 2		
	Gear 3 Teeth = _____	A / 3			B / 3		
	Gear 4 Teeth = _____	A / 4			B / 4		
	Largest gear = 5 Teeth = _____	A / 5			B / 5		

Turn the bicycle upside down. Set the gears at each combination. Turn the rear wheel so that the 0 degree position is at a bar of the frame. Apply friction to the rear wheel while you turn the pedal one 360 degree rotation. Record how many degrees of turning are made by the wheel. Remember that one complete rotation is 360 degrees. Record the results and determine the number of turns in decimal notation by dividing the degrees of turn by 360 degrees per turn..

		Largest Pedal Gear = A			Smallest Pedal Gear = B		
		Gears Connected	Degees of Turn	Decimal Equivalent	Gears Connected	Degrees of Turn	Decimal Equivalent
Wheel Gears	Smallest Gear 1	A / 1			B / 1		
	Gear 2	A / 2			B / 2		
	Gear 3	A / 3			B / 3		
	Gear 4	A / 4			B / 4		
	Largest Gear 5	A / 5			B / 5		

TOOTH to TOOTH

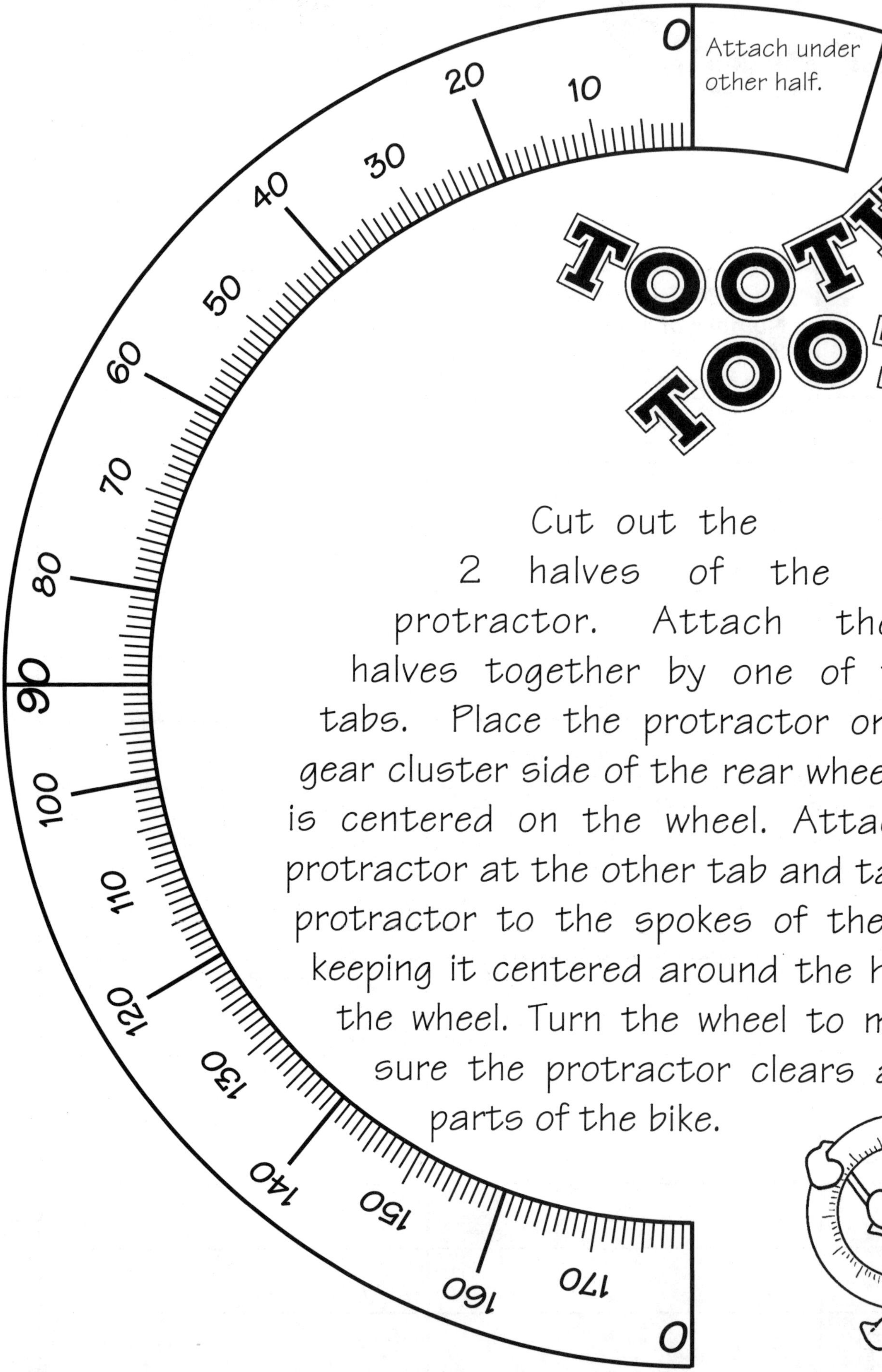

Cut out the 2 halves of the protractor. Attach the halves together by one of the tabs. Place the protractor on the gear cluster side of the rear wheel so it is centered on the wheel. Attach the protractor at the other tab and tape the protractor to the spokes of the wheel keeping it centered around the hub of the wheel. Turn the wheel to make sure the protractor clears all parts of the bike.

TOOTH to TOOTH

Designer Gears

Topic Area
Gears

Introductory Statement
In this activity students will draw designs using a design toy such as a Spirograph®. Students will discover several methods of predicting how a design will look before drawing it by using the number of teeth on the gears.

Math
Using rational numbers
- fractions
- factors
- prime factors
- multiples
- least common multiple
- greatest common factor
- ratios

Science
Physical science
- Simple machines

Math/Science Process Skills
Observing
Predicting
Collecting and recording data
Interpreting data
Generalizing
Applying

Materials
For each group:
- design toy such as a Spirograph®
- 3" X 3" self-stick removable notes

For teacher preparation:
- cover stock or tag board
- thumbtacks
- tape

Key Questions
1. Can you find a numerical pattern to explain the visual patterns formed by the gears?
2. Can you apply this numerical pattern to a design before you draw it?

Background Information
A design toy such as a Spirograph® is based on planetary gear systems, where one gear rotates while it revolves inside or around a stationary gear. This type of gear arrangement gets its name from its similarity to the planets rotating as they revolve around the sun. By tracing one point of the rotating gear, a design is drawn. Different designs are drawn by using different combinations of gears or by placing the tracing point at different positions. With all designs, a general pattern emerges. Every rotation of the moving gear produces a *petal* if it revolves inside the stationary gear. However, if it revolves around the outside of the stationary gear, it produces an *indent.* If the number of rotations can be calculated, the design can be predicted.

To complete a design, the drawing position must come back to the exact place it started. For this to happen, the moving gear must complete a rotation and a revolution simultaneously. The moving gear will have rotated or spun a whole number of times, and it also will have revolved or circled around or within the stationary gear a whole number of times.

The following information will help to find number patterns that will explain the visual patterns drawn by the gears.

- The size of the gear must be used. The size of the gear, in this case, is the number of teeth the gear has. This number is usually printed on the gear.
- The number of rotations multiplied by the number of teeth on the moving gear gives the number of teeth used in the pattern.
- The number of teeth on the stationary gear multiplied by the number of revolutions gives the number of teeth used in the pattern.
- The number of teeth used in rotations and revolutions is the same since the teeth are meshed.

Since the number of teeth involved in the rotations and the revolutions must be equal, and the design is complete when the number of rotations and revolutions are completed simultaneously, the design may be predicted by listing the multiples of the number of teeth on the gears. Listing the multiples of the moving gear's number of teeth tells the number of teeth used after each complete rotation. Listing the multiples of the stationary gear's number of teeth, tells the number of teeth used after each complete revolution. The first time these two lists of multiples have a common member (*least common multiple*) is the first time the gears will be in the original position and the design is complete. The number of rotations done can be determined by the position of the common member in the listing of the moving gear's teeth multiples. The number of revolutions done can be determined by the position of the common member in the listing of the multiples of the stationary gear's teeth.

Another method of predicting the design is to determine the *prime factorization* of the two gears used. Students will determine the prime factors for the number of teeth on both gears. The product of these factors in the union of these two sets is the number of teeth used to complete the design. (Each factor must be viewed as a separate number and must be used to find the product.For instance, if the number 2 appears three times, it must be multipled three times (...2 x 2 x 2...) and not just once). Dividing this number by the moving gear size and stationary gear size respectively will give you the number or rotations and revolutions. The two prime factorizations can also be put into a two circle Venn diagram. The product of the numbers appearing in the intersection of the two circles is the *greatest common factor* of the two gear sizes. The product of the factors in the rotation set, not in the intersection, is the number of revolutions. The product of the factors in the revolution set, not in the intersections, is the number of rotations.

A final way to predict a design is to use a ratio of the gear sizes. Ratios are the way all gear sets are described and this is a natural extension of students' development of the understanding of gears. The moving gear size is compared to a stationary gear size in a ratio and then reduced to simplest terms. The numbers in this reduced ratio are the revolutions and rotations needed to make the design. Although this method often provides the easiest way to a solution, it also lacks depth of meaning for students. It is presented last in this activity so students will see its connection to the other two methods.

Management

1. Allow four to five sessions or periods for the completion of this activity: one to two periods to gain practice using the design toy, complete the drawings, and record the numbers of rotations and revolutions; another period to list the multiples and discover this application; one period to work on prime factorization and Venn diagrams; and a final period to reduce ratios and bring closure to the activity.
2. To complete the entire activity, it is assumed that students have some familiarity with multiples, least common multiples, factors, and prime numbers. If the teacher does not feel that all these skills are appropriate for the students, the activity may be stopped after either of the first two methods which were described in the *Background Information*.
3. Because many students will have had no experience with design toys, it is best to allow them time to explore and practice using them before beginning the activity. Also, if the teacher has not used them, some time should be taken before the activity, to become familiar with their operation.
4. The procedure is written as if each group of six students has a design toy. This is generally the exception rather than the rule. If only one kit is available, set up a station where two groups of three students can work. In each case, one student will draw the design , the second student will count rotations, and the third will count revolutions. One group of three will use the larger ring for drawing, the other group of three will use the smaller ring. After having the initial discussion and basic instructions, have students work through the activity, making designs and recording the number of rotations and revolutions.
5. Students may have some difficulty differentiating between a *rotation* (a gear spinning) and a *revolution* (a gear traveling in a circular path). Before starting the activity, the teacher must spend some time clarifying this difference. Be certain to let group members decide who will count rotations and who will count revolutions. It is very difficult for one student to do both.
6. Before beginning the activity, the teacher should make the *Designer Drawing Spaces.* Copy the sheet onto cover stock, or glue copies to tag board. Cut the sheet into four sections. Place thumbtacks up from the bottom at the cross-hairs. Tape the tack heads to the bottom of the cards. Make enough for each group of three students if the class will be using the design toys at the same time. Four will be enough for a center.
7. To use the drawing spaces, the students will place a 3" X 3" removable, sticky note in the position over the tag board cutouts, pushing the thumb-tacks through the note. They will place the holes found in the ring gear onto the thumbtacks and draw the design. The drawings are then removed and mounted onto the record sheet.
8. Because this activity has so many student pages, it is best to distribute them as progress is made throughout the activity so as not to overwhelm the students.

Procedure

Parts 1 and 2: Rotations, Revolutions, and Least Common Multiples

1. Explain to the students how a design toy works. Distribute toys, removable note pads, drawing spaces, and allow students to explore.
2. Discuss *Key Question* with students.
3. Describe and demonstrate the differences between rotations and revolutions.
4. Have students complete the 12 drawings following the instructions on the record sheet. If students are working in groups of six, each group of three could make the drawing for one ring. They could then share information and drawings with another group which used the other ring to complete the activity.
5. Have students complete the worksheets where they list the multiples of the ring and gears by following the instructions.
6. Direct students to refer to their drawings and their multiples charts for discussing patterns in the charts and patterns in the drawings.

Part 3: Prime Factorization and Venn Diagrams

7. Discuss the *Key Question.*
8. Make certain that students have their designs which they drew from *Parts 1* and *2* and distribute the handouts for the prime factorization and Venn diagrams.
9. Have students follow the instructions to complete the worksheets where they find the prime factorization of the gear sizes. It may be easier for students to draw factor trees of the numbers of teeth on the gears and

rings before attempting to place the numbers in the Venn diagrams. If students are not familiar with Venn diagrams, some instruction may be required from the teacher. When using ring size 96, there are instances when no number is written in one of the circles of the Venn diagram other than in its intersection with the other circle. Students often erroneously assume that no number equals zero. To prevent this in these instances, it may be helpful to have students write the numeral 1 (a factor of all numbers) in the circle even though it is not a prime factor.

10. Have students refer to their drawings and their pages on prime factorization for discussion.

Part 4: Ratios and Conclusion

11. Once again, ask the *Key Question* and distribute the final handout. Make certain that students have their drawings and previous handouts from making comparisons of data.
12. Have students follow the instructions to complete the worksheet on ratios. (Students might be encouraged to use the product of the intersection of prime factorizations to find the greatest common factor.)

Discussion

Parts 1 and 2:

1. What visual clue in the drawings tells you how many rotations were made? [The number of petals equals the number of rotations.]
2. What pattern did you find between the numbers on the left hand side of the chart and the numbers on the right hand side of the chart? [Revolutions x ring teeth = rotations x gear teeth.]
3. From using the multiples chart, is there a way one could predict the number of petals a design would have if one knew the gear size? [Student should conclude that the first time the multiple listings match (*least common multiple*) is the place where the drawing is complete. The number of rotations needed to get to this position is the number of petals in the design.]

Part 3:

4. Is there a way one could predict the number of petals a design would have using prime factorization? Explain. [The product of the prime factors of only the ring size gives the number of rotations, hence the number of petals.]
5. What do you think the product of prime factors for the gear size represents? [The number of revolutions to complete the design]
6. What does the product of the prime factors of both the ring size and gear size represent? [The total number of gear teeth involved in making a complete design; it is the least common multiple for the two numbers]
7. What does the product of prime factors common to the ring size and gear size represent? [the greatest common factor of the two numbers]
8. Compare the results from *Parts 1* and *2* to the results recorded on the Venn diagrams. Explain how they are related.

Part 4:

9. How can you use the simplified ratios to predict the number of petals in a design? [Using the simplified ratio, the number corresponding to the gear tells how many revolutions were made; the number corresponding to the ring tells how many rotations, or petals, were made. The students might be encouraged to note that the gear/ring size relationship is a revolution/rotation relationship similar to teeth and turn ratios of other gears.]
10. How can you predict what a design would look like that uses gear 63 with ring 96? [Students may solve the problem using any of the methods discovered or a combination of methods.] How many petals would the design have? [32 petals]
11. Discuss the advantages the calculation method has over actually drawing the solution. [speed, accuracy, not material bound]

Extensions

1. Have students explore how revolving on the exterior side of a gear differs from revolving on the interior.
2. Allow students to explore and determine if their generalizations and methods work on the assorted pieces in the set.
3. Let students draw their own designs and give them to other students to determine the gears used.

Major Conceptual Components

- Energy comes in many forms.
- Mechanical energy is the energy due to the position or the movement of something.
- A machine is a device for multiplying force or changing the direction of force.
- Wheels and axles are turning levers.
- Gears are toothed wheels.
- Gears can be used to transfer force from one part of a machine to another.

Designer Drawing Space

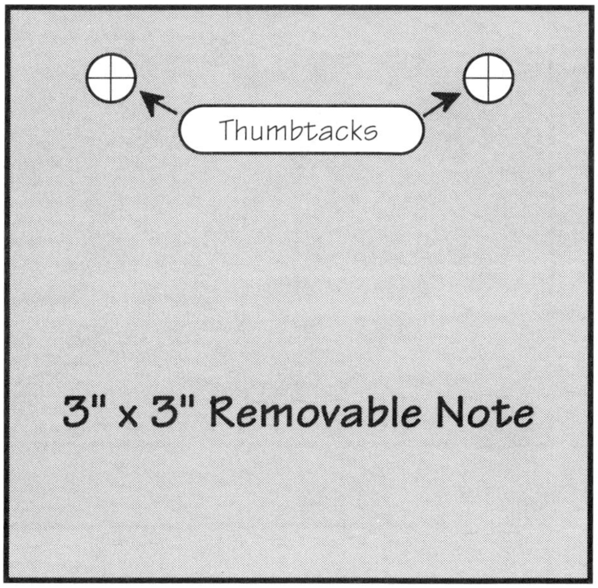

Designer Drawing Space

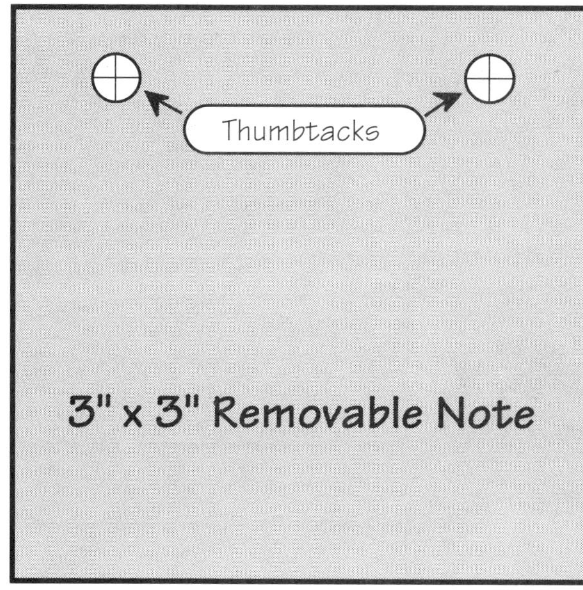

Designer Drawing Space

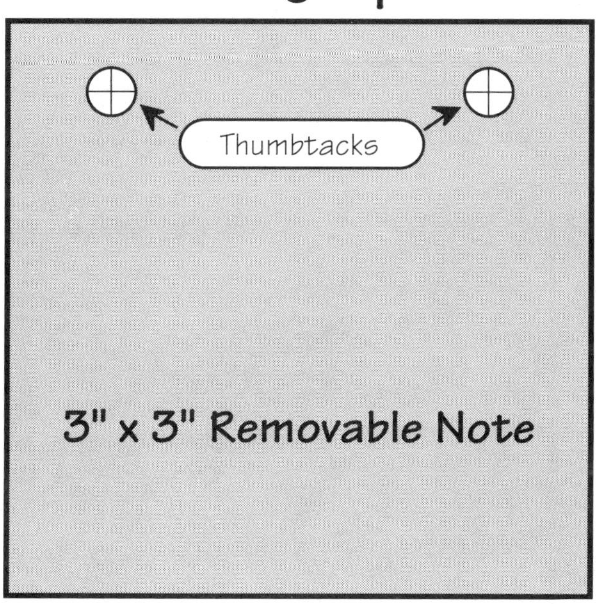

Designer Drawing Space

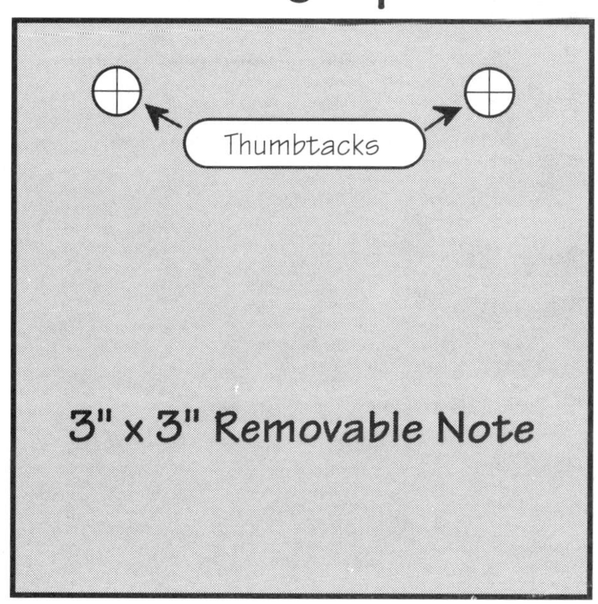

DESIGNER GEARS

Ring 96 with Wheel 24 *or* Ring 105 with Wheel 30	Ring 96 with Wheel 32 *or* Ring 105 with Wheel 45	Ring 96 with Wheel 48 *or* Ring 105 with Wheel 56
Ring 96 with Wheel 60 *or* Ring 105 with Wheel 60	Ring 96 with Wheel 72 *or* Ring 105 with Wheel 63	Ring 96 with Wheel 80 *or* Ring 105 with Wheel 84

ROTATIONS/REVOLUTIONS

Can you find a numerical pattern to explain the visual patterns formed by designer gears?

Use the gears and rings specified in the chart.

1. Place a pencil in hole 1 of the gear and place the gear at the starting position 1 of the ring.

2. Roll the gear around the ring until the lines match up, keeping track of the number of revolutions and rotations.

   When the gear rolls by or up to the starting position on the ring, it is a revolution.

   When the starting point (hole 1) of the gear comes in contact with the ring, it is a rotation.

3. Follow steps 1 and 2 for each of the gear sets listed on the charts. Record the data on the charts, and on each drawing label the gear and ring used. Save each drawing.

Revolutions around Ring	Teeth on Ring	Teeth on Gear	Rotations of Gear
	96	24	
	96	32	
	96	48	
	96	60	
	96	72	
	96	80	

Revolutions around Ring	Teeth on Ring	Teeth on Gear	Rotations of Gear
	105	30	
	105	45	
	105	56	
	105	60	
	105	63	
	105	84	

What visual clue is there on the drawings to tell you how many rotations were made?

What pattern can you find between the numbers on the left half of the chart and their corresponding numbers on the right side of the chart?

LEAST COMMON MULTIPLES

Ring 96

Each time a gear rotates it uses each of the teeth on its circumference. The 24 tooth gear uses 24 teeth on the first rotation. By the second rotation, it has used 48 teeth (2x24). On the third rotation, 72 teeth have been used (3x24).

Each time a gear revolves inside a ring it passes the number of teeth on the circumference of the inside of the ring. On the 96 tooth ring it passes 96 teeth on the first revolution, 192 on the second, and 288 on the third.

When a design pattern is completed, the rotations and revolutions are completed at the same spot. The number of teeth used during rotations is equal to the number of teeth passed in revolving.

Complete the chart below by completing the total number of teeth passed in revolutions and the total teeth used in rotations for each gear size. Then for each gear size circle the first time the total teeth of rotation match the total teeth of revolution.

96 Tooth Ring

Revolution	1	2	3	4	5	6	7	8
Total Teeth	96	192	288					

Gear Size	Total Teeth for Each Rotation														
	1	2	3	4	5	6	7	8	9	10	11	12	13	14	15
24	24	48	72	(96)											
32															
48															
60															
72															
80															

How can you use a listing of the multiples of the gear and ring teeth to predict what a design pattern will look like?

LEAST COMMON MULTIPLES

Ring 105

Each time a gear rotates it uses each of the teeth on its circumference. The 30 tooth gear uses 30 teeth on the first rotation. By the second rotation, it has used 60 teeth (2x30). On the third rotation, 90 teeth have been used (3x30).

Each time a gear revolves inside a ring it passes the number of teeth on the circumference of the inside of the ring. On the 105 tooth ring it passes 105 teeth on the first revolution, 210 on the second, and so forth.

When a design pattern is completed, the rotations and revolutions are completed at the same spot. The number of teeth used during rotations is equal to the number of teeth passed in revolving.

Complete the chart below by completing the total number of teeth passed in revolutions and the total teeth used in rotations for each gear size. Then for each gear size circle the first time the total teeth of rotation match the total teeth of revolution.

105 Tooth Ring

Revolution	1	2	3	4	5	6	7	8
Total Teeth	105	210						

Gear Size	Total Teeth for Each Rotation														
	1	2	3	4	5	6	7	8	9	10	11	12	13	14	15
30	30	60	90												
45															
56															
60															
63															
84															

How can you use a listing of the multiples of the gear and ring teeth to predict what a design pattern will look like?

Prime Factorization

Ring 96

Listing the multiples of the *gear* and *ring* sizes is one way to predict how many rotations and revolutions will be required to complete a *design pattern*. Other number patterns show up if the *gear* and *ring* sizes are broken down into their prime factorizations.

1. Write the prime factorization of the *gear* and *ring* sizes used in each *design*.
2. Circle the prime factors that are shared in each of the prime factorizations.
3. List this set of common prime factors in the intersection of the 2 circles on the *Venn* diagrams.
4. Put the remaining prime factors in their respective circles on the diagrams.
5. Find the product of each region of the *diagram* and the product of all the prime factors.

Prime Factors

96 = (3x2x2)x2x2x2 ____ 5x(3x2x2) = 60

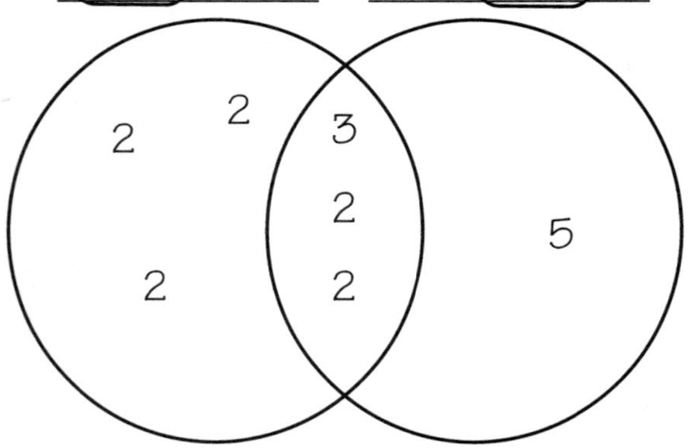

Product of prime factors of only 96: 8

Product of prime factors common to 96 and 60: 12

Product of prime factors of only 60: 5

Product of prime factors of 96 and 60: 480

Prime Factors

96 = ______ ______ = 24

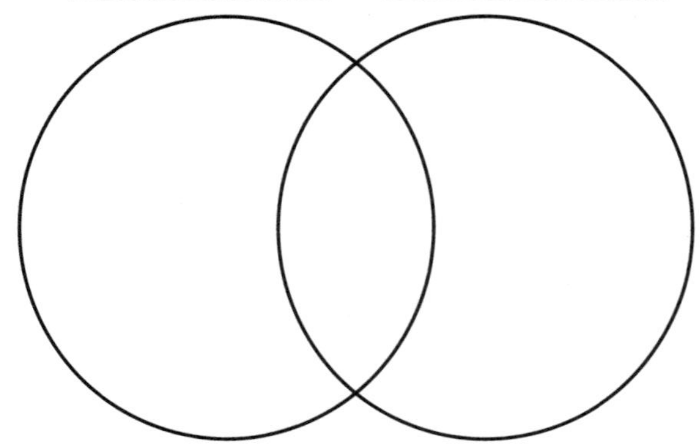

Product of prime factors of only 96: ______

Product of prime factors common to 96 and 24: ______

Product of prime factors of only 24: ______

Product of prime factors of 96 and 24: ______

How can you use a prime factorization of the *gear* and *ring* sizes to predict what a design pattern will look like?

Prime Factorization

Ring 96

Prime Factors

96 = ______ ______ = 32

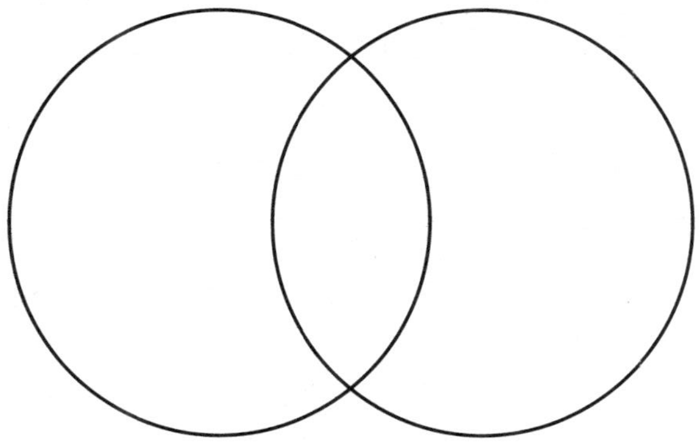

Product of prime factors of only 96: ______
Product of prime factors common to 96 and 32: ______
Product of prime factors of only 32: ______
Product of prime factors of 96 and 32: ______

Prime Factors

96 = ______ ______ = 48

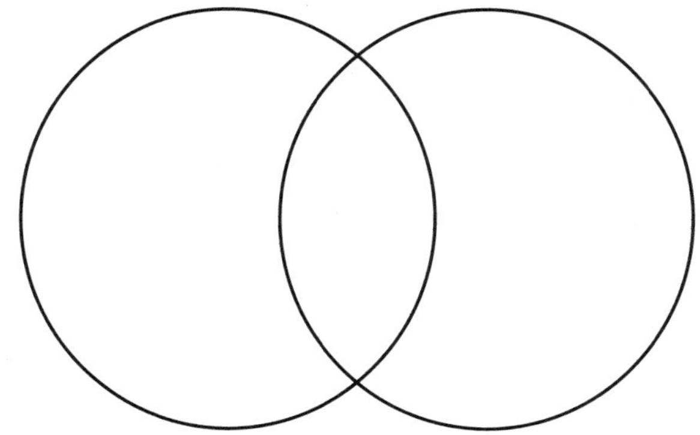

Product of prime factors of only 96: ______
Product of prime factors common to 96 and 48: ______
Product of prime factors of only 48: ______
Product of prime factors of 96 and 48: ______

Prime Factors

96 = ______ ______ = 72

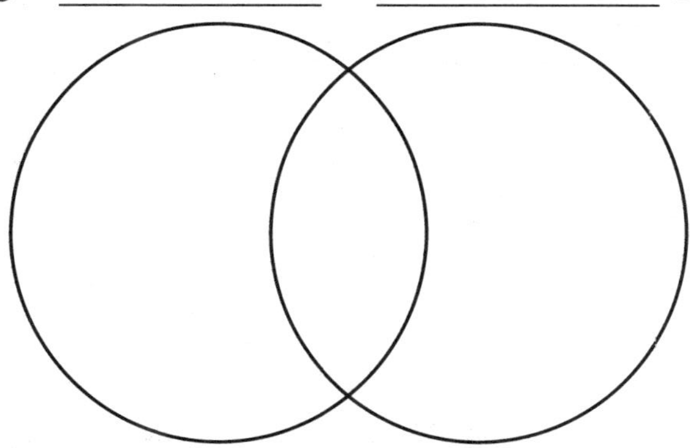

Product of prime factors of only 96: ______
Product of prime factors common to 96 and 72: ______
Product of prime factors of only 72: ______
Product of prime factors of 96 and 72: ______

Prime Factors

96 = ______ ______ = 80

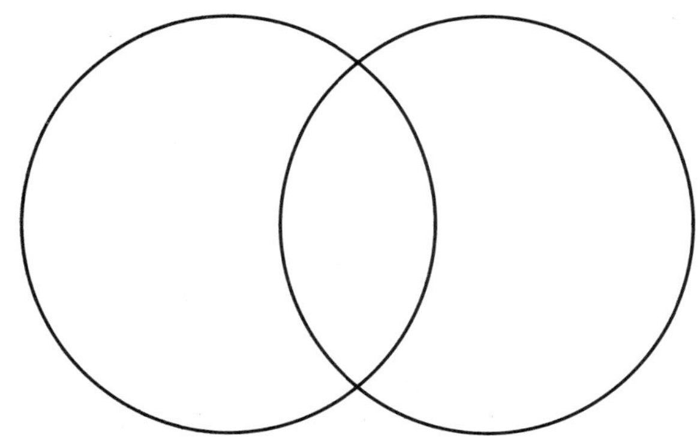

Product of prime factors of only 96: ______
Product of prime factors common to 96 and 80: ______
Product of prime factors of only 80: ______
Product of prime factors of 96 and 80: ______

Prime Factorization

Ring 105

Listing the multiples of the gear and ring sizes is one way to predict how many rotations and revolutions will be required to complete a design pattern. Other number patterns show up if the gear and ring sizes are broken down into their prime factorizations.

1. Write the prime factorization of the gear and ring sizes used in each design.
2. Circle the prime factors that are shared in each of the prime factorizations.
3. List this set of common prime factors in the intersection of the 2 circles on the Venn diagrams.
4. Put the remaining prime factors in their respective circles on the diagrams.
5. Find the product of each region of the diagram and the product of all the prime factors.

Prime Factors

105 = 7 x (5 x 3) (5 x 3) x 2 = 30

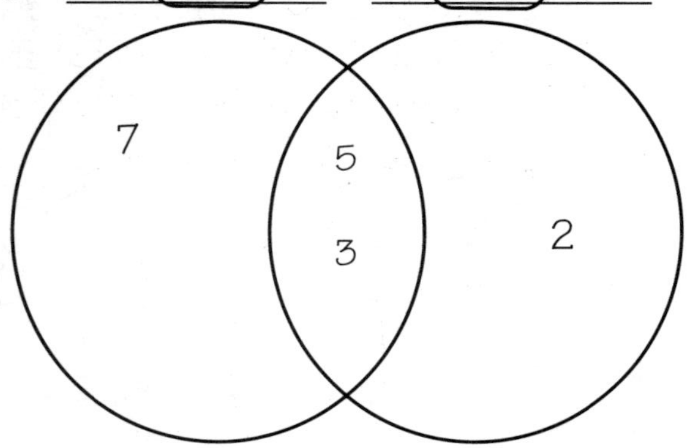

Product of prime factors of only 105: 7
Product of prime factors common to 105 and 30: 15
Product of prime factors of only 30: 2
Product of prime factors of 105 and 30: 210

Prime Factors

105 = ______ ______ = 45

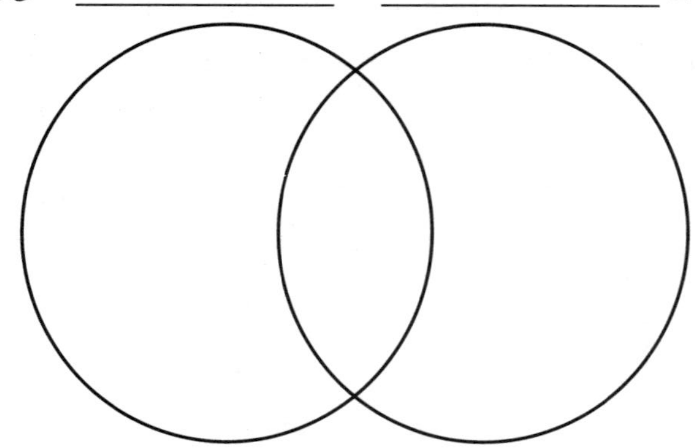

Product of prime factors of only 105: ______
Product of prime factors common to 105 and 45: ______
Product of prime factors of only 45: ______
Product of prime factors of 105 and 45: ______

How can you use a prime factorization of the gear and ring sizes to predict what a design pattern will look like?

Prime Factorization

Ring 105

Prime Factors

105 = ______ ______ = 56

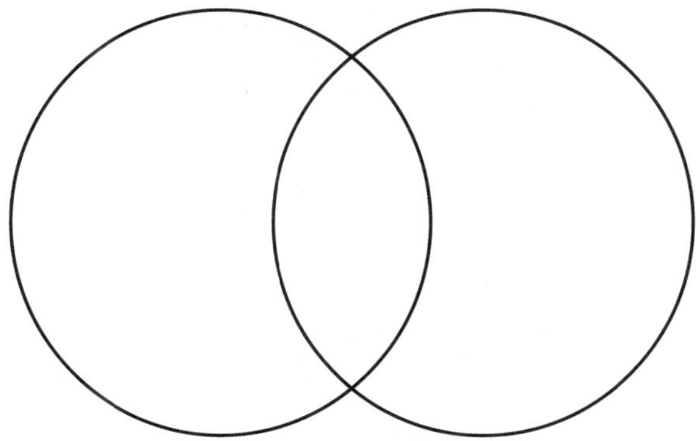

Product of prime factors of only 105: ______
Product of prime factors common to 105 and 56: ______
Product of prime factors of only 56: ______
Product of prime factors of 105 and 56: ______

Prime Factors

105 = ______ ______ = 60

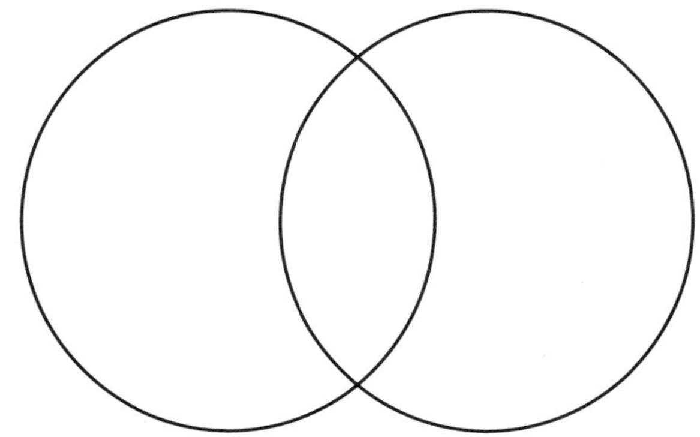

Product of prime factors of only 105: ______
Product of prime factors common to 105 and 60: ______
Product of prime factors of only 60: ______
Product of prime factors of 105 and 60: ______

Prime Factors

105 = ______ ______ = 63

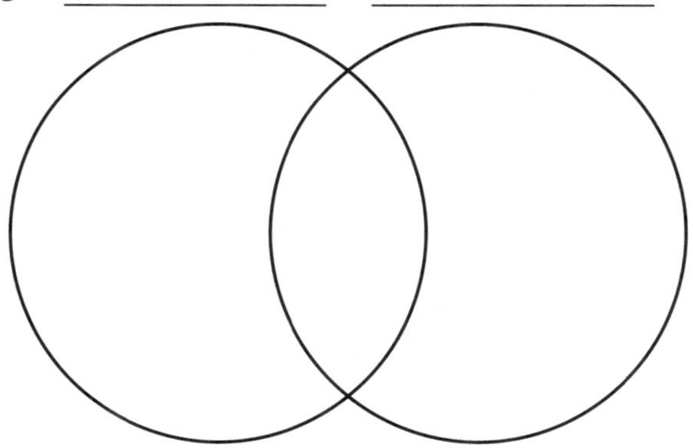

Product of prime factors of only 105: ______
Product of prime factors common to 105 and 63: ______
Product of prime factors of only 63: ______
Product of prime factors of 105 and 63:

Prime Factors

105 = ______ ______ = 84

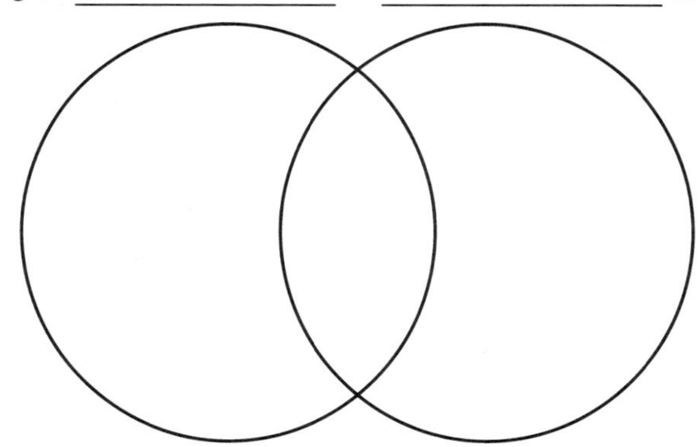

Product of prime factors of only 105: ______
Product of prime factors common to 105 and 84: ______
Product of prime factors of only 84: ______
Product of prime factors of 105 and 84:

Ratios

How can you use a gear ratio to predict what a design pattern will look like?

Designer gears are a form of a planetary gear system where one gear rotates while it revolves within another gear.

All gear combinations are describe by tooth ratios. These ratios provide information about how much each gear turns. Use the ratios on the chart to discover what they reveal about the turns.

Find the greatest common factor for each tooth ratio, then simplfy each ratio.

Tooth Ratio Gear / Ring	Greatest Common Factor	Simplified Ratio
24/96		
32/96		
48/96		
60/96		
72/96		
80/96		
30/105		
45/105		
56/105		
60/105		
63/105		
84/105		

April 4

Dear Journal,

It was one of those beautiful spring days when everything is right with the world. A group of the neighborhood kids got together at the vacant lot to fix up the tree house which was built in a large oak tree. We went to work with hammers and nailed down the loose boards.

Mucky came along looking very dejected. He has never been able to climb the ladder up to the first level and so was feeling left out again.

We all took a break and sat down to talk to Mucky. When we were trying to make him feel better, we hit on the idea of putting in an elevator. (Bet that was an uplifting idea for Mucky!) While everyone else went back to work on the house, Mucky and I headed back to the machine shop for some parts.

I took a quick look in the encyclopedia to get any added information I might need about pulleys.

> A pulley is a wheel used to support, guide, and transfer force from a moving rope, belt or chain. A system of pulleys forms a simple machine that multiplies force at the expense of distance.

Since the encyclopedia did not seem to give me any new ideas, I grabbed the box of machine parts labeled "Pulleys." I carried the box, Mucky got some rope, and we headed back to the tree house.

We tied one pulley to a large limb above the tree house. We slipped the rope through and tied a loop on one end for Mucky to sit in. Mucky stepped into the loop while we climbed up to the tree house. As soon as we were at the top, I grabbed the rope and jumped out of the tree house expecting to drop to the ground, but nothing happened. I just hung there. Several other friends grabbed onto my end of the rope and hung there with me. Still Mucky wouldn't budge. I heard a sign of relief from Greedy Gravity who still had Mucky in her grasp.

We figured more was better, so we decided to try three more pulleys. We took a long rope and rigged up a pulley system with Mucky on one end and the rest of us holding onto the rope up on the tree house. I jumped onto the rope…nothing happened. A second friend climbed on and suddenly the rope started to drop. As we went down, Mucky started to move up. But not all was so great. Behind our screams of excitement,

I could hear the laugh of Fiendish Friction. He was up to something. When we finally reached the ground, Mucky was up, but less than half way. An elevator would be no good if it could not get Mucky to the top.

We lowered Mucky to the ground. Since it was getting dark, we all headed for home and figured we would solve this problem later.

April 11

Dear Journal,

As I was headed to the vacant lot today, I saw Handy planting some flowers. I discussed the problem of the pulleys with him to see if he had any advice. He took me back to the shop and told me I had two problems.

First, he told me that the pulleys were very old and dirty which was causing a problem with their efficiency. He opened the encyclopedia and told me to read about efficiency.

> Efficiency is the percent of work put into a machine that is converted to actual work. Efficiency can be calculated using the following ratio:
>
> Efficiency = Work Out / Work In

Handy said that my friend and I were doing a lot more work than just moving Mucky. We were moving and pulling at gunk and dirt in the pulleys. This friction in the pulleys was using up a lot of our work so there was not as much left to lift Mucky. Handy got me a can of grease and some oil to make my machine more efficient.

Handy told me I had a second problem because I had not read the encyclopedia carefully. I had missed the key phrase, "A system of pulleys forms a simple machine that multiplies force at the expense of distance." Handy told me the pulleys multiplied our force so we could lift Mucky, but we did not move him very far. If we wanted to move him farther, we would have to move much farther.

I raced to Mucky's to get him so we could go the tree house. When we got to the vacant lot, we cleaned, greased, and oiled the pulleys. Mucky climbed into his loop, and we climbed up the tree. I grabbed the rope, jumped out of the tree house, and quickly glided to the ground (slick, huh? Another pun, I'm good!), and Mucky rose. I anchored the rope at the ground level, scrambled up the tree again, and grabbed the rope, jumped, and lifted Mucky some more. I kept doing this until he was to the tree house.

We enjoyed the rest of the day. Mucky was the most pleased and he smiled with the joy of having escaped Greedy Gravity's clutches. At the end of the day, Mucky climbed into the loop, holding all our tools. This time, as he dropped, he gave us turns to be lifted up to the tree house.

April 12

Dear Journal,

I was so excited about yesterday's discovery that as soon as I got home from school, I ran over to see Handy. He was sitting on his porch enjoying the beautiful afternoon. I told him all about what happened yesterday. I asked him if he could explain how the pulleys worked. He told me that by connecting the pulleys the way we did, that many ropes were holding Mucky's weight. Each rope held up a fraction of Mucky's weight. The rope that I held had to be pulled with the force of that fraction of Mucky's weight in order to move. I must have weighed just a little bit more than that fraction of Mucky's weight because my weight was able to start the rope moving slowly. The pulley system multiplied my force. It had a mechanical advantage.

Handy said to imagine that we had used six pulleys so there were six sections of rope that ran through those pulleys. The six ropes would hold up Mucky's weight rather than just one. Each rope would hold one-sixth of his weight. That meant that I would have to pull with the force of one-sixth of his weight. The pulley system would multiply my force by six times giving it a mechanical advantage of six. This whole thing with machines is starting to make some sense.

Handy went on to explain that since six ropes held Mucky up, each rope had to be shortened one meter to get Mucky to rise one meter. Six ropes shortened one meter each means the rope would have to be pulled six meters to get Mucky to rise one meter. Now I understand why I had to pull the rope so many times.

I am beginning to realize that the concept that you multiply force at the expense of distance works for all machines. With the pulleys, I could lift Mucky who is six times my weight, but I had to pull the rope six times as far as he went. Next time I have a problem with machines, I am going to remember this.

Meet the Equalizer

Topic Area
Pulleys

Introductory Statement
After watching one student beat a team of students in a tug-of-war, students will measure the amount of pull they can generate. They will then determine how many pulleys and support lines are needed to win a tug-of-war with the other members of their group.

Math
Measuring
  force
Using whole number operations
Using and developing formulae
Using calculators

Science
Physical science
  Simple machines

Math/Science Process Skills
Observing
Collecting and recording data
Interpreting data
Applying and generalizing
Comparing

Materials
50 feet of 1/4 inch synthetic rope
Bathroom scale
2 broom handles or 2 36-inch pieces of 1" doweling
Smooth stationary pole (flagpole, basketball pole, etc.)
Tape measure
Optional: gloves

Key Question
How would you set up a system of pulleys or lines of support so that you could win a tug-of-war against the rest of your group?

Background Information

No machine can put out more energy than is put into it. No machine can create energy; it can only transfer it from one place to another, or transform it from one form to another.

A pulley is a type of first-class lever that consists of a wheel as a fulcrum and a rope as the lever arms. The rope which is attached to a resistance passes over the wheel (the fulcrum) and ends with the other end serving as the source for the application of the effort.

A pulley can change the direction of a force. An elevator is an example of a single pulley in which the direction of its movement is changed. When the elevator moves upward, the counterweight moves downward. But when the elevator moves downward, the counterweight moves upward. With a single pulley, the resistance moves the same distance as the effort.

A system of pulleys not only changes the direction of the effort's force, but can also amplify that effort. However, no new energy has been created, the amplification of the effort was at the expense of distance. The amount by which a pulley system amplifies the effort to raise the resistance depends on how many lines of support the system contains. The amplification is equal to the number of lines of support within the pulley system. (For more information about lines of support see the *Background Information* for the activity *Pulley Power*).

In this activity, the action of a pulley can be demonstrated by a tug-of-war in which one child offsets the weight and strength of two or three other children. To do this, you will need **at least** three supporting lines. When an even number of lines of support is needed, one end of a rope is tied to a stationary pole, such as a flagpole. The rope is then looped around both a broomstick and the flagpole several times. The other end of the rope is tied to second dowel or broomstick. When an odd number of lines of support is desired, the one end of the rope is tied to the first broomstick instead of the stationary pole.

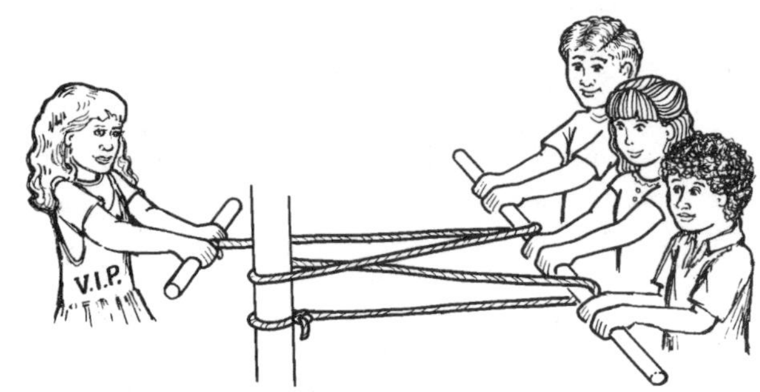

The individual "Very Intense Puller (VIP)" pulls on the dowel on the end of the rope, while a team of two or three people pull at the same time on the broomstick that is looped several times by the rope. The individual should be able to win this tug-of-war because of the mechanical advantage of the pulley system. Each line of support increases the mechanical advantage.

As more lines of support are added, the amount of friction multiples until the advantage of another rope is counteracted by the amount of friction it generates.

Management

1. Students must have an understanding of how the amount of lines in a pulley system increase its mechanical advantage. The activity *Pulley Power* may be done to develop this understanding.
2. The smoothest rope possible should be used in the pulley system to reduce the friction that occurs during the tug-of-war.

3. When students are measuring force, they should only pull against the scale. Encourage students not to look at the scale when they are pulling against it. There is a great temptation to get a larger amount of force by squeezing the scale, instead of just pulling it.
4. For safety, students should hold the dowel close to the rope. This will cause less stress on the dowel, reducing the likelihood of the dowel snapping. Students will need to take care not to get rope burns. You might want to have gloves available for them to wear for protection.
5. Before beginning this activity, the teacher will want to find an appropriate place to put the pulley system and will want to set it up once to become familiar with the set-up procedure.
6. Before class, the teacher may find it appropriate to prepare the scale and its rope harness as shown on the first student page.
7. Allow two 45-minute periods for this activity.
8. This activity is designed to be completed in small groups of 3-4 students. It may be completed as a whole class activity by using the demonstration team's numbers to complete all calculations.

Procedure

1. Tell the class you are going to have a tug-of-war. Ask for volunteers: one individual to be the "Very Intense Puller (VIP)" and two for the opposing team. No measuring of force is done at this time as this procedure serves as a demonstration to show students that the opposing team can easily defeat the VIP in a traditional tug-of-war.
2. Have the tug-of-war.
3. Talk about the outcome of the war and discuss the *Key Question*.
4. Divide the class into small groups of 3-4 students. Have each group choose an individual to be the VIP with the rest of the group forming the opposing team.
5. Have the groups write their names in the appropriate place on the first chart. Direct individual group members to predict the force with which they think they will be able to pull the rope.
6. Have groups take turns measuring each individual's pulling force against the scale as illustrated and explained on the first handout.
7. Using their force measurements, have the students predict the fewest number of lines of support necessary to allow the VIP to win in a tug-of-war. Students may complete the calculations to draw these conclusions.
8. Have each group go out to the tug-of-war site and test its calculated predictions. Students are to adjust the system, increasing or decreasing the number of lines by making more or less loops around the team's broomstick. The number of lines of support should match the predicted amount. Before starting the contest, the students should mark the starting position of the team's broomstick, and the VIP's.
9. If the opposing team's predicted number of lines of support was not enough to allow the VIP to win, add one line of support at a time and see if the VIP gains the advantage. The group is to continue adding lines of support until the VIP is able to win. The group records the number of lines needed, and completes the actual force calculations.
10. After the VIP has won, students need to measure and record the distances the broomsticks have moved from their starting marks.
11. Students should be asked to make conclusions about the number of lines and the distances moved by completing the sheet.

Discussion

1. How did the actual amount of lines of support differ from the amounts you predicted and what could account for this difference? [The predicted amount of lines of support is theoretical (ideal). As the amount of lines increase, the friction increases and the chance for getting lines tangled increases. Friction is not taken into account in the predicted calculations.]
2. What could you do to cut down on the friction? [grease the rope, grease the pole, use fewer pulleys]
3. What relationship is there between the lines of support and the distances moved? [The ratio of the VIP's distance to the team's distance is equal to the number of lines of support. Students should notice that the distance ratio is more closely related to theory than is force. This is an excellent chance for students to see that force is multiplied at the expense of distance.]

Extension

Set up a demonstration with a block and tackle and have one student pull a car or bus.

Major Conceptual Components

- Energy cannot be created or destroyed.
- The amount of energy or work put into a machine will come out of the machine in some form.
- Mechanical advantage is a comparison of force produced by a machine compared to the force applied to the machine.
- Ideal mechanical advantage is the calculated value determined from theoretcial understanding. It does not take into account the effect of forces such as friction.
- Actual mechanical advantage is the comparison of the actual generated force to the applied force.
- A machine is a device for multiplying force or changing the direction of force.
- Pulleys can be used to increase the effort force and/or change the direction of the force.
- The effort needed to raise the resistance depends on the number of lines of support the pulley system contains.

MEET the EQUALIZER

Directions:

1. Tie a rope tightly around a doorknob so it is unable to slip off.
2. Make a loop out of the other end of the rope. Check and make sure the loop is large enough to fit a bathroom scale. Tie a knot in the loop that will not slip. Make sure the door is securely closed.
3. Predict the amount of force each person will be able to pull and record it on the chart below.
4. Have the "Very Intense Puller (VIP)" and each member of the team of opponents pull against the bathroom scale that is held by the rope tied to the doorknob while another team member reads the amount of force. Record the force for each person on the chart.

	Amount of Pull	
VIP	Prediction	Actual
Opposing Team	Prediction	Actual
	Team Total	

Use the information from the chart to determine how many lines of support will be needed to make the tug-of-war slightly favorable for the VIP. The equation and inequality below will help you determine the number of lines that will be needed in the pulley system.

Total Force of Opposing Team	<	VIP with Pulleys	=	VIP's Force	x	Mechanical Advantage	Lines of Support

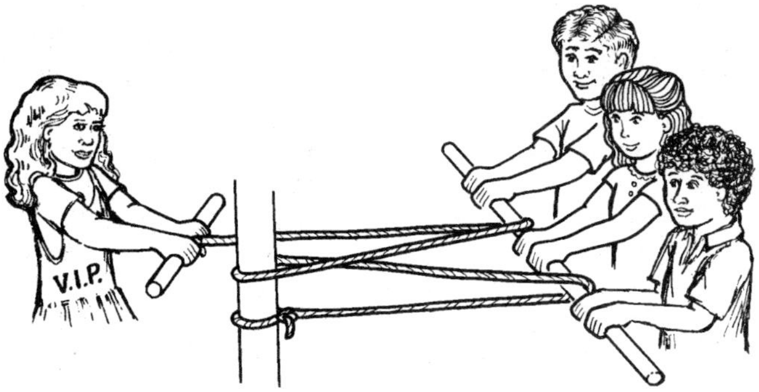

Now go out to the tug-a-war area and test your prediction.

1. Loop the rope around the group's stick and the stationary pole so there are the number of supporting lines your group predicted.
2. Have the VIP and the opposing team take their sticks and pull them so the rope becomes taut. Mark the starting positions of the sticks for both sides on the ground.
3. Start the tug-a-war. If the VIP does not win, adjust the rope to make one more line of support and try it again. Continue adding lines of support until the VIP wins. Before starting each new trial, remark the starting positions.
4. When the war has been won by the VIP, measure the distance the VIP's and the team's sticks have moved from their starting positions. Record the distances on the chart below.
5. Record the number of lines of support needed for the VIP to win and calculate the actual force the VIP generated.

Distance Team Moved	Distance VIP Moved

Total Force of Opposing Team		VIP with Pulleys	=	VIP's Force	x	Mechanical Advantage	Lines of Support Needed to Win
	<						

6. How many times farther did the VIP move than the team?

7. What relationship is there between the mechanical advantage and the distance moved?

8. Why did the VIP have to multiply his/her force to be able to win?

Block and Tackle

Topic Area
Ideal mechanical advantage

Introductory Statement
Students will construct a block and tackle and determine its ideal mechanical advantage. They will then vary the configuration of the block and tackle to determine how that affects the ideal mechanical advantage.

Math
Constructing a block and tackle model
Measuring
Tabulating information
Computing mechanical advantage
Averaging

Science
Physical science
  Simple machines

Math/Science Process Skills
Observing
Collecting and recording data
Interpreting data
Comparing
Inferring
Drawing conclusions
Predicting

Materials
For each group of two students:
- 1 20 cm x 40 cm piece of corrugated cardboard
- 3 large paper clips
- 1 rubber band, approx. 7cm in length
- meter stick or metric tape
- heavy polyester thread
  - 1 m length for mechanical advantages of 1-3
  - 2 m length for mechanical advantages of 4-6
- tape

Key Question
What determines the ideal mechanical advantage of a block and tackle?

Background Information
A block and tackle is a system of ropes and pulleys used for lifting heavy loads. It makes it possible to lift a *heavy weight* over a relatively *short distance* by applying a *small force* over a relatively *long distance.*

The system typically consists of one rope wound around two separate sets of pulleys. The upper set is attached to a support and the lower set to the load that is to be moved. Applying sufficient force to the rope raises the lower set of pulleys and the load. In this activity the paper clips perform the role of pulleys and the thread that of the rope.

Students need to understand the meaning of ideal mechancial advantage and actual mechanical advantage. The *ideal mechanical advantage* assumes that the system is frictionless. The *actual mechanical advantage* is found in the real world where some of the force is used to overcome the friction in the system.

Strange as it may seem, therefore, students will determine the approximate *ideal mechanical advantage* of various systems in this investigation by comparing the distance the force moves to the distance the load is lifted. The relative distances are affected so little by friction (which causes some stretching of the thread) that the effect is minimal. Instead of measuring the relative distances moved by the force and the load, if we would actually measure the relative forces, the result would be the *actual mechanical advantage.*

In this investigation, if in one of the systems the force moves a distance of 2 cm to lift the load 1 cm, the ideal mechanical advantage is 2. Similarly, if the force moves 12 cm to lift the load 2 cm, the ideal mechanical advantage is 6.

The *ideal mechanical advantage* is equal to the number of rope sections supporting the load. In this activity that will range from 1 to 6.

Management
1. The corrugated cardboard can be obtained from discarded boxes. After use, it can become a permanent part of the "equipment" for science. Pieces should be cut to size ahead of time.
2. It is important to perform the activity in groups of two, so that the students are maximally involved in the construction of the apparatus and threading of the pulleys systems. Both boys and girls should take turns in threading the pulleys to enhance learning.

Procedure
1. Explain the meaning of *ideal* and *actual mechanical advantage.*
2. Form groups of two, if possible, and distribute the materials.
3. Guide students through the process of constructing the system. Have the whole class construct and investigate the first system together to become familiar with the procedure.
4. Instruct student groups to complete investigating the remaining systems as illustrated on the student's page.
5. Ask students to generalize the results.
6. Discuss the generalizations and the reasons why errors in measurement occurred.

Discussion
1. What is the difference between *ideal* and *actual mechanical advantage*?

2. What are quick ways to determine the *ideal mechanical advantage* of a system?
3. Where can you find applications of block and tackle systems?

Curriculum Correlations

Social Studies:

In what occupations in our history did the block and tackle play a significant role? (sailing ships, machine shops, factories)

Home Links

1. Determine where block and tackle systems are used in your community.
2. Have students find out whether or not parents use a block and tackle in their work; have someone who uses them come to explain the task(s) for which they are used.

Major Conceptual Components

- Energy cannot be created or destroyed.
- The amount of energy or work put into a machine will come out of the machine in some form.
- Mechanical advantage is a comparison of force produced by a machine compared to the force applied to the machine.
- Ideal mechanical advantage is the calculated value determined from theoretical understanding. It does not take into account the effect of forces such as friction.
- Actual mechanical advantage is the comparison of the actual generated force to the applied force.
- A machine is a device for multiplying force or changing the direction of force.
- Pulleys can be used to increase the effort force and/or change the direction of the force.
- The effort needed to raise the resistance depends on the number of lines of support the pulley system contains.

BLOCK AND TACKLE

Construction of Block and Tackle

1. Tape the scale at the left onto a 20 cm X 40 cm piece of cardboard.
2. Cut two notches at the bottom of the cardboard for attaching the rubber band to the system.
3. Construct the block and tackle as shown below.

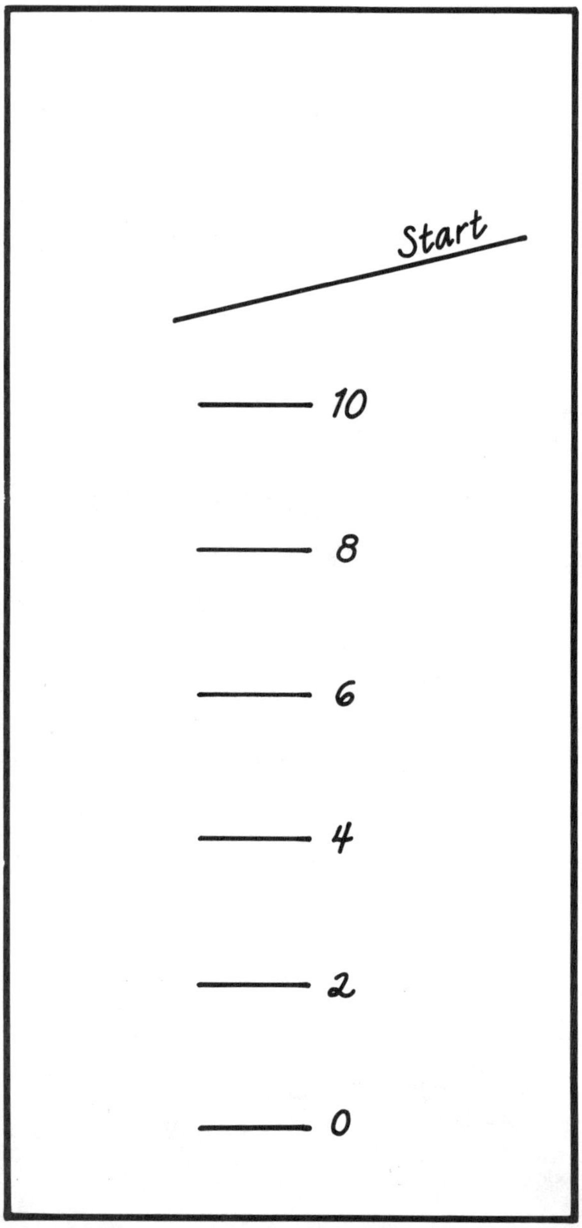

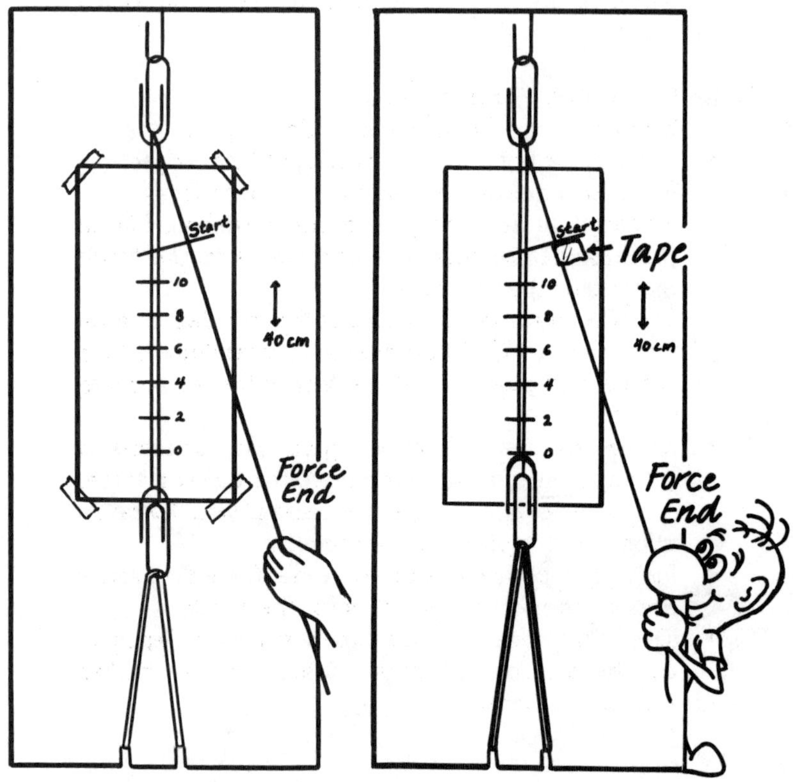

Using the Block and Tackle

1. Bring the top of the paper clip to the zero mark on the scale by pulling on the force end of the thread.
2. Fasten a piece of tape around the force end of the thread at the point where it crosses the "start" line.
3. For the first test, raise the paper clip to the 2 centimeter mark.
4. Measure the distance the tape on the force end of the thread had to be moved and record the result.
5. Continue in the same manner for the other distances.

BLOCK AND TACKLE

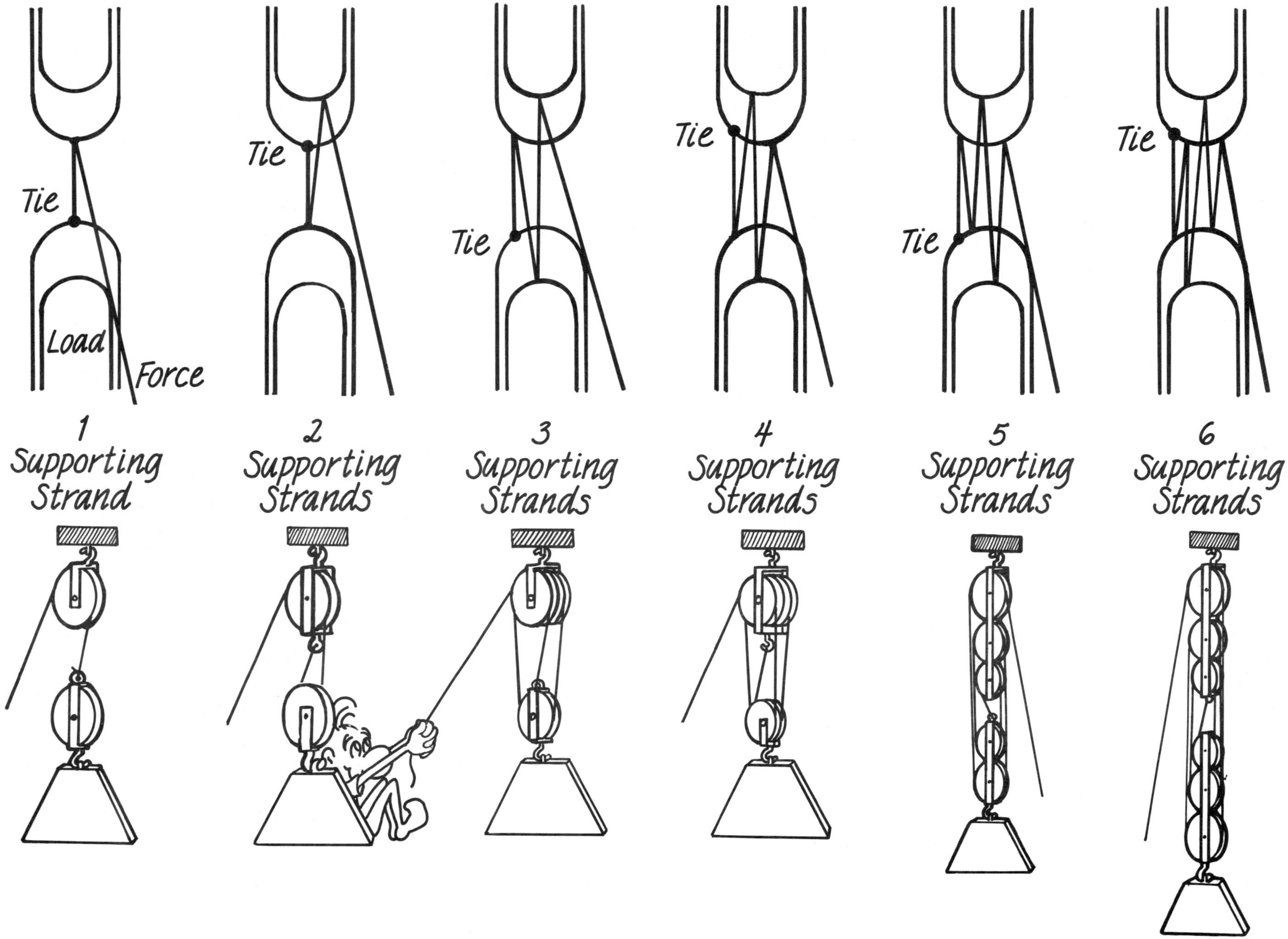

BLOCK AND TACKLE

Record the information from your investigations in the following tables. The mechanical advantage is found by dividing the distance the force moves by the distance the load moves.

1 Load Support Strand

Distance Load Moves	2 cm	4 cm	6 cm	8 cm	10 cm	
Distance Force Moves						Average Mechanical Advantage
Mechanical Advantage						

2 Load Support Strands

Distance Load Moves	2 cm	4 cm	6 cm	8 cm	10 cm	
Distance Force Moves						Average Mechanical Advantage
Mechanical Advantage						

3 Load Support Strands

Distance Load Moves	2 cm	4 cm	6 cm	8 cm	10 cm	
Distance Force Moves						Average Mechanical Advantage
Mechanical Advantage						

4 Load Support Strands

Distance Load Moves	2 cm	4 cm	6 cm	8 cm	10 cm	
Distance Force Moves						Average Mechanical Advantage
Mechanical Advantage						

BLOCK AND TACKLE

5 Load Support Strands

Distance Load Moves	2 cm	4 cm	6 cm	8 cm	10 cm	
Distance Force Moves						Average Mechanical Advantage
Mechanical Advantage						

6 Load Support Strands

Distance Load Moves	2 cm	4 cm	6 cm	8 cm	10 cm	
Distance Force Moves						Average Mechanical Advantage
Mechanical Advantage						

Please enter the computed mechanical advantage into this table.

Mechanical Advantage Summary

Number of Support Strands	Distance Moved by Load					Average Mechanical Advantage
	2 cm	4 cm	6 cm	8 cm	10 cm	
1						
2						
3						
4						
5						
6						

What conclusions can you draw? ______________________________

____

____

____

Pulley Power

Topic Area
Pulleys

Introductory Statement
Students will construct a pulley system and examine such factors as number of supporting lines, mechanical advantage, and friction.

Math
Measurement
  length
Graphing
Averaging
Using calculators
Estimating

Science
Physical science
  Simple machines

Math/Science Process Skills
Observing
Predicting
Inferring
Collecting and recording data
Interpreting data
Applying and generalizing

Materials
Per group:
  15 paper clips
  thin rubber band
  meter stick or ring stand
  monofilament fishing line
  masses–200-500 grams (see *Management*)
  rulers
  1/2 lb margarine tub

Key Question
How does a pulley system make it easier to lift heavy objects?

Background Information

A pulley system is a simple machine typically consisting of one or more pulleys and a line to pull the load.

A pulley may be fixed or movable. The *fixed pulley* does not move: It helps us only by changing the direction of the force. The *movable pulley* moves along the rope. It allows a gain in force but increases the distance over which the force must be applied. In a single movable pulley, two sections of the rope support the pulley. One end is attached to something stationary, the other end is lifted up. Only half as much effort is needed to raise the resistance or weight. However, the effort must now move about twice as far as the resistance.

Here are diagrams of supporting lines on different pulley systems. A *supporting line* must hold up part of the weight of the object. If the line merely changes the direction of the pull, it is not a support line. When the final line of a pulley system is pulled in an upward direction, it is a support line. When it is pulled in a downward direction, it is merely changing the direction of the pull and does not support any additional weight.

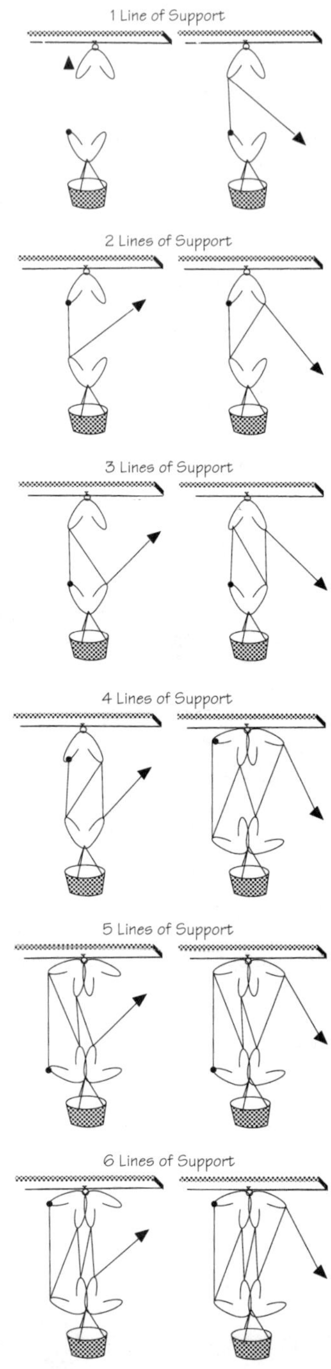

When one end of the rope is attached to the load, or movable pulley, there will be an odd number of support lines. When one end is attached to a stationary object, or fixed pulley, an even number of support lines are used.

Ideally the weight of an object divided by the number of supporting lines will give the force needed to lift the object with the help of the pulley system. The number of support lines times the distance the object needs to be lifted gives the total length of line needed to be pulled to raise the object to a given height. For example, if it takes 100 units of force to lift an object 150 centimeters straight up, then with one pulley and two supporting lines, it should take 50 units of force, but you will have to pull the rope 300 centimeters.

The force of friction is very evident in pulley systems. The ideal mechanical advantage is often significantly better than the actual. Generally, the more movable pulleys used, the less force will be needed. However, the reduction of force is not uniform with the addition of each pulley. Increasing the number of pulleys also increases the friction from rope contact, and within the pulleys. At some point, adding more pulleys may increase the friction to a point that the actual mechanical advantage decreases.

A pulley is usually constructed of a wheel and axle. This is done to decrease the effects of friction. In this activity, a pulley is simply the wire of a paper clip. To keep friction to a minimum, it is necessary to use monofilament fishing line.

Management

1. This activity has three parts. If the class period is 40-50 in length, two to three periods will be needed.
2. Before doing the activity, the teacher should prepare the tubs so they all have a uniform mass of 200-500 grams. If commercial mass sets are not available, pennies, ceramic tiles, or any other heavy items with consistent masses may be used.
3. The pulley system may be supported by a meter stick placed between two desks or two sturdy chairs. If ring-stands are available, these may be used instead. The system should be elevated as high as possible for easier access.
4. All groups must use rubber bands of the same size.
5. Before doing the activity, the teacher should assemble a pulley system for demonstration. Refer to the instructions on *Pulley System Assembly* handout. Provide a copy of these instructions for each group of students.
6. The paper clips used as pulleys need to be opened far enough so the lines do not get tangled easily.
7. In this activity, students will not be able to mathmatically determine that the weight of the load divided by the number of lines of support equals the force needed to lift an object using a pulley system. However, they will be able to experience the use of less force (and see this through less stretch of the rubber band) with the addition of more lines of support.
8. Students will need to know how to tie an overhand knot to secure the fishing line.

Procedure

Part 1:

1. Discuss the *Key Question.*
2. Demonstrate the teacher-made pulley system. Discuss supporting lines and how to tell the difference between them and lines that merely change the direction of the pull. Distribute *Count On My Support.* Have students complete this handout and use it to assess their comprehension of lines of support.
3. Distribute *What A Line!* to reinforce what has been demonstrated and determine what will be tested. Students may work in pairs.

Part 2:

1. Divide students into groups of 3-5 and distribute *Pulley System Assembly* and the materials to each group.
2. Have each group build a pulley assembly.
3. Distribute *Stretched to the Limit.*
4. Have students measure the rubber band at rest and record the length in the appropriate column on the chart. Instruct the students to measure this length to the nearest millimeter but record in centimeter units (Ex: 5.3 cm). This measurement should stay the same for each set of pulleys.
5. Ask students to predict the amount of net stretch for the rubber band using one line of support. Remind them that the net stretch is the difference between the length of the rubber band at rest and the stretched rubber band. Referring to the instructions for stringing the pulleys, have students follow directions and lift the tub. They should measure the stretched length of the rubber band again to the nearest millimeter but record in centimeter units, calculate the net stretch, and compare their predictions with the actual stretch.
6. Repeat procedure 5 for the other numbers of lines of support.

Part 3:

1. Distribute worksheets *Power of the Mean* and *Sketch the Stretch.*
2. Have groups share their results while each student records the class results on *Power of the Mean.*
3. Direct students to find the average net stretch for the class and make a bar graph from the data found on worksheet *Sketch the Stretch.*
4. Have students complete and discuss the generalization at the bottom of *Sketch the Stretch.*

Discussion

1. What can you say about the mechanical advantage of the pulleys?
2. When less force is required to lift a load, what is sacrificed? [distance] Explain.
3. What factors affect the efficiency of the pulley? [friction, the ropes tangling, uneven setup can cause more pull at one part of the system instead of an even pull]
4. Where would you find pulleys in everyday life? [curtain pulls, block and tackles, flag poles]

Extensions

1. Use a string instead of monofilament line and observe the effects of friction.
2. Use a spring scale to measure effort instead of observing the rubber band's stretch.
3. Bring in a block and tackle to demonstrate the mechanical advantage gained by a pulley system.
4. Have students find actual examples of pulleys and either bring them in or draw pictures of them.

Major Conceptual Components

- Energy cannot be created or destroyed.
- The amount of energy or work put into a machine will come out of the machine in some form.
- Mechanical advantage is a comparison of force produced by a machine compared to the force applied to the machine.
- Ideal mechanical advantage is the calculated value determined from theoretical understanding. It does not take into account the effect of forces such as friction.
- Actual mechanical advantage is the comparison of the actual generated force to the applied force.
- A machine is a device for multiplying forces orchanging the direction of forces.
- Pulleys can be used to increase force and/or change the direction of the force.
- The effort needed to raise the resistance depends on the number of lines of support the pulley system contains.

COUNT ON MY SUPPORT

Write the number of lines of support for each drawing.

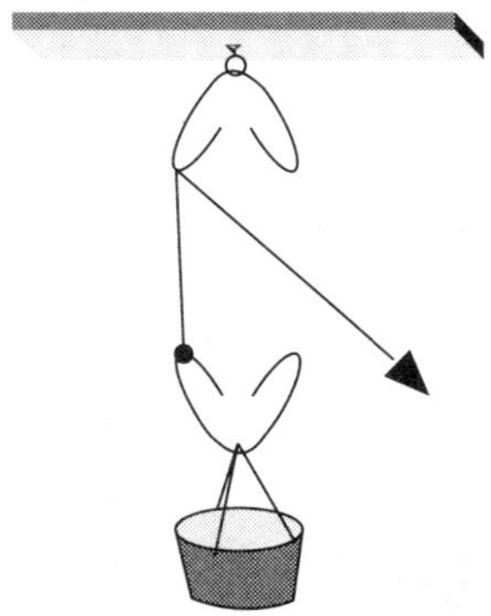

1. ____________

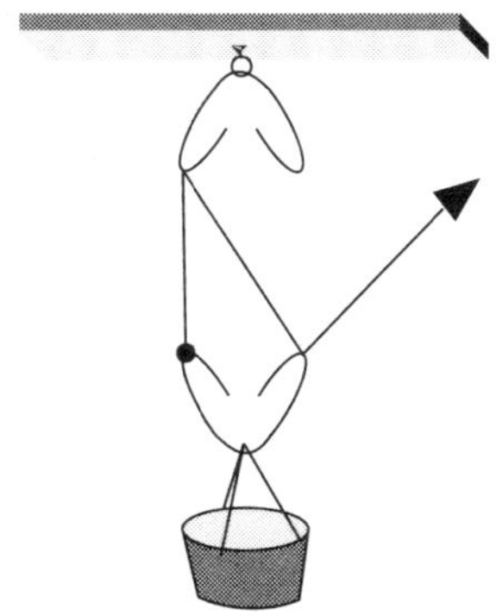

2. ____________

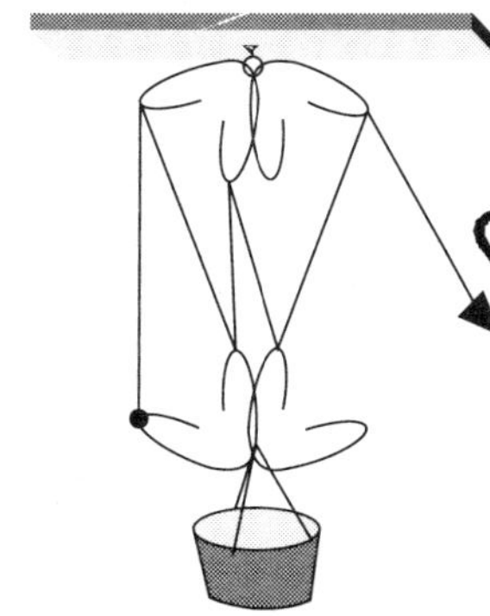

3. ____________

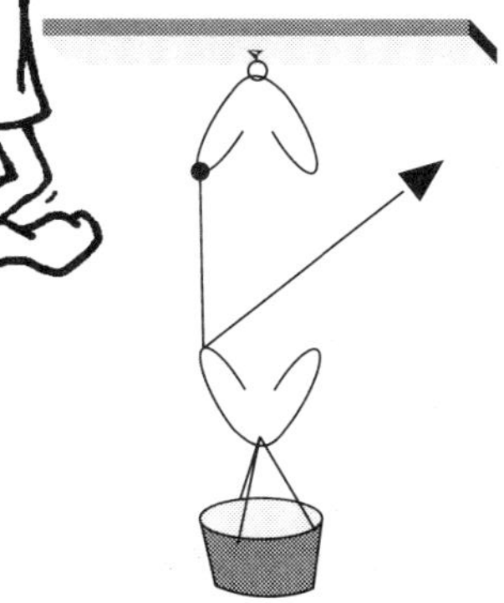

4. ____________

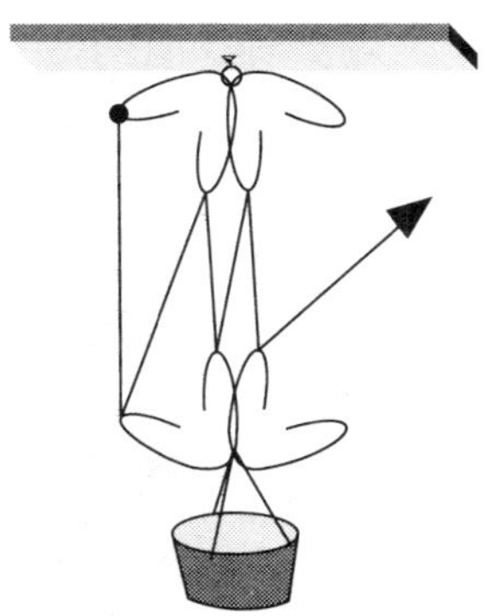

5. ____________

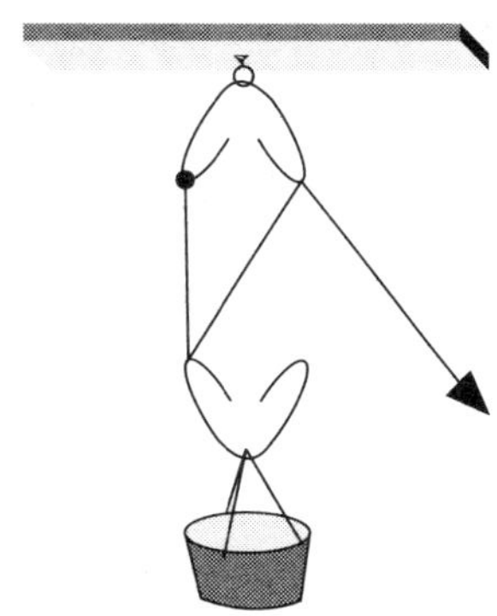

6. ____________

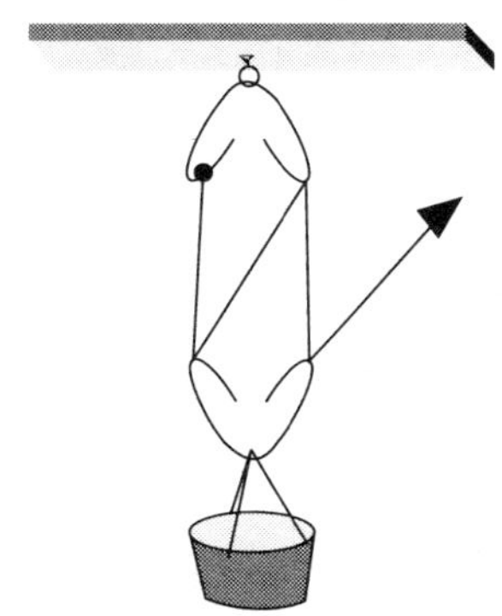

7. ____________

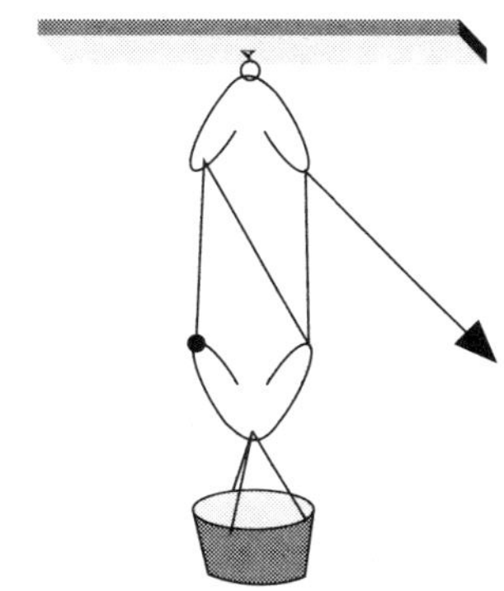

8. ____________

In all the drawings, where is the string attached that show an even number of lines of support?

In all the drawings, where is the string attached that show an odd number of lines of support?

What a Line!

Draw a pulley system for *each* of the number of lines of support listed below.

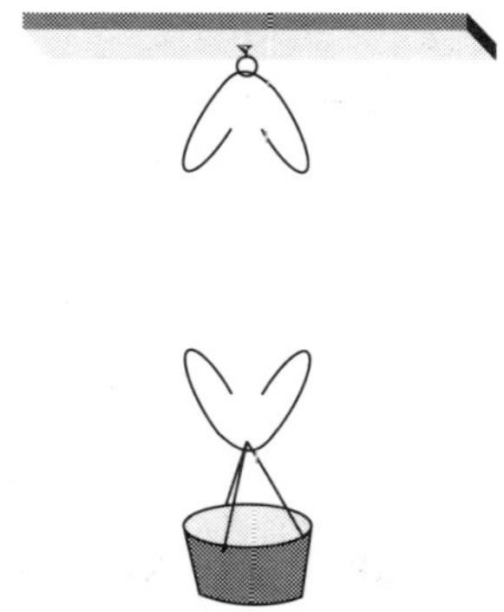

1 Line of Support

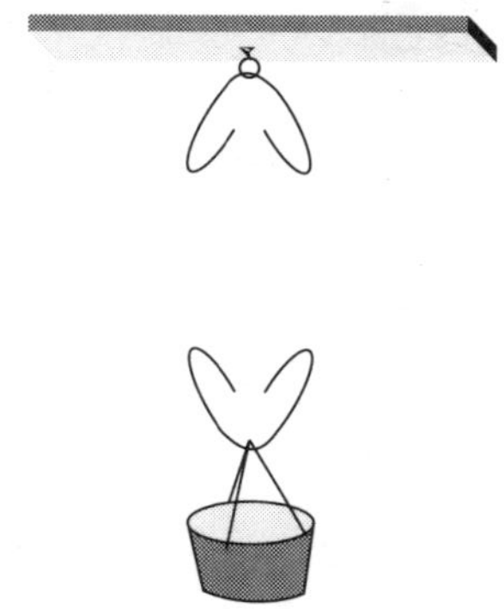

2 Lines of Support

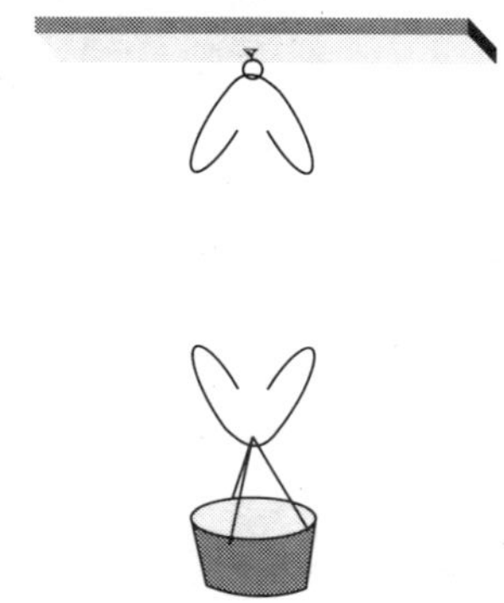

3 Lines of Support

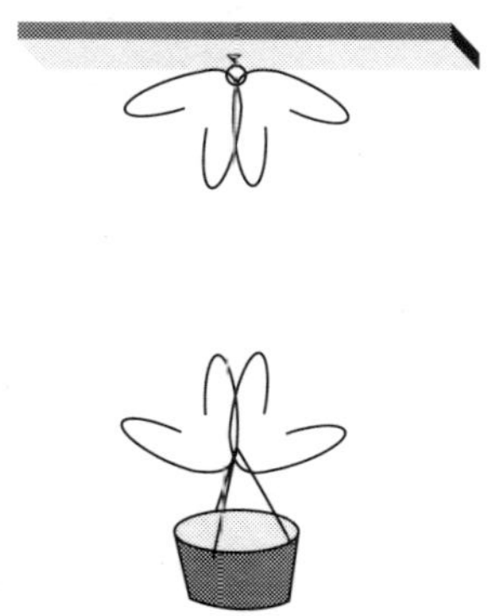

4 Lines of Support

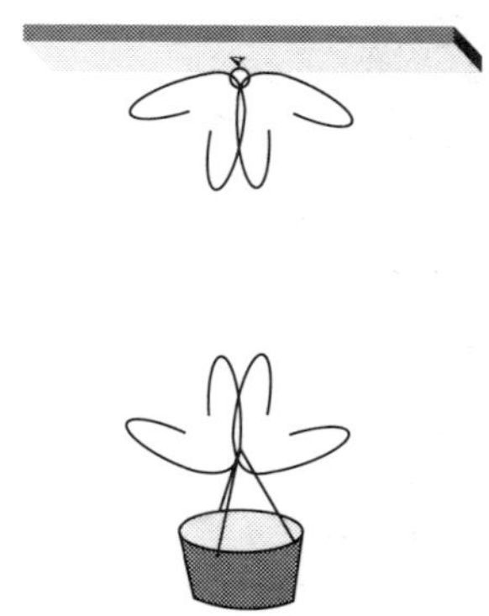

5 Lines of Support

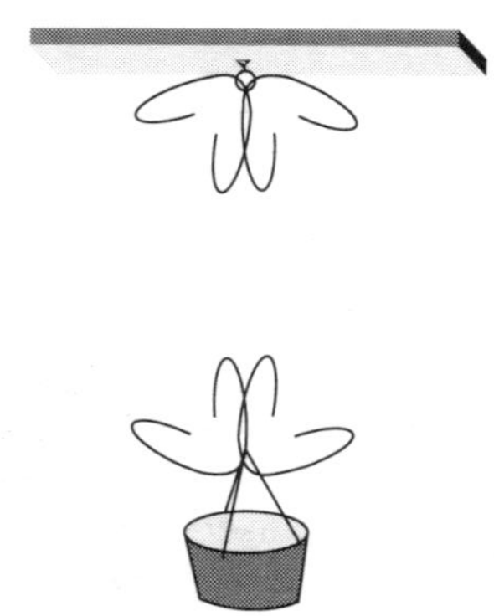

6 Lines of Support

Can you draw another way to make a system that will have the *same amount* of lines as *each* of the drawings above?

PULLEY POWER

SYSTEM ASSEMBLY

1. Use 10 paper clips to make a tub hanger as shown in the picture.

2. With a piece of fishing line, make a loop big enough to go around the width of a meter stick. Secure the loop with a overhand knot and slide the loop onto the middle of the meter stick.

3. Set each end of the meter stick on different desks to suspended it.

4. Spread four paper clips so each forms two open loops.

5. Hang two of the bent paper clips from the meter stick through the loop in the fishing line. The loops of these clips will be the fixed pulleys.

6. Insert the other two bent paper clips through the top paper clip of the tub hanger. The loops of these clips will be the movable pulleys.

7. Make a small loop with an overhand knot at one end of one meter of fishing line.

8. Tie the rubber band to the other end of the fishing line.

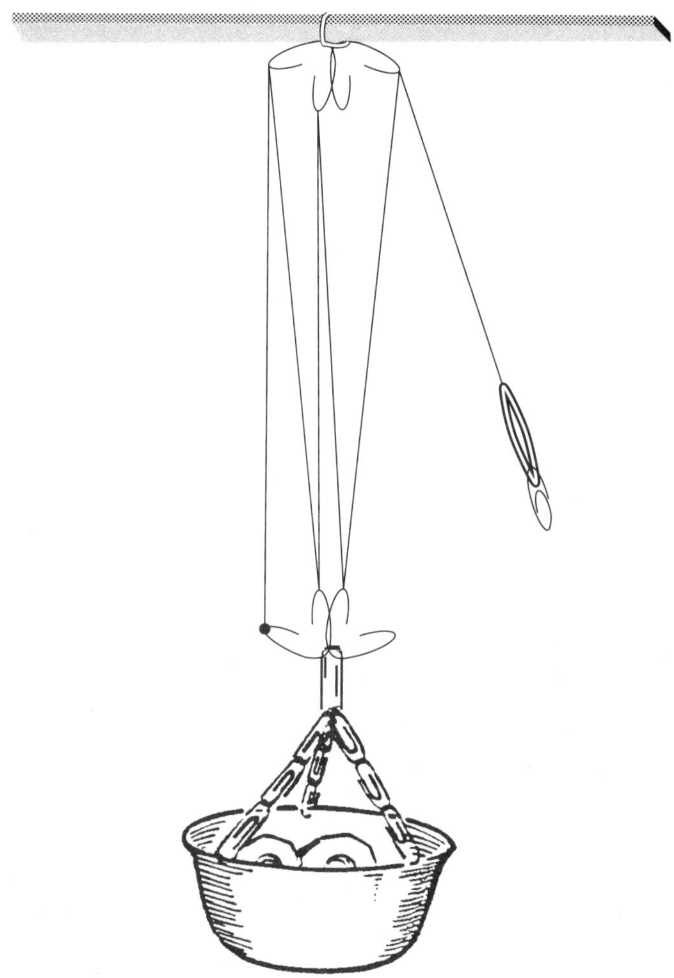

9. Attach a paper clip through the rubber band to pull the line.

10. The loop on the other end of the line will secure the line to the pulley system.

STRETCHED TO THE LIMIT

How will the length of the stretched rubber band change as the number of lines of support change?

1. Assemble your pulley system for 1 line of support.
2. Measure and record the length of the rubber band at rest.
3. Estimate and record how much the rubber band will strecth when you lift the tub.
4. As the tub is lifted, measure and record the length of the rubber band and calculate the net stretch.
5. Complete the chart by following steps two to four for each number of lines.

Number of Lines of Support	Estimate of Net Stretch	Rubber Band Stretched (cm)	− Rubber Band at Rest (cm)	= Net Stretch (cm)
1				
2				
3				
4				
5				
6				

What relationship do you see between the stretch in the rubber band and the number of supporting lines?

POWER OF THE MEAN

Record the centimeters of net stretch for each group in the class. Calculate the total and average net stretch for the class

Group	Number of Lines of Support					
	1	2	3	4	5	6
1	cm	cm	cm	cm	cm	cm
2	cm	cm	cm	cm	cm	cm
3	cm	cm	cm	cm	cm	cm
4	cm	cm	cm	cm	cm	cm
5	cm	cm	cm	cm	cm	cm
6	cm	cm	cm	cm	cm	cm
7	cm	cm	cm	cm	cm	cm
8	cm	cm	cm	cm	cm	cm
Total	cm	cm	cm	cm	cm	cm
Average	cm	cm	cm	cm	cm	cm

SKETCH THE STRETCH

Construct a bar graph using the data recorded on *Power of the Mean*. Then make a line across all the groups' bars to represent the average for that amount of supporting lines.

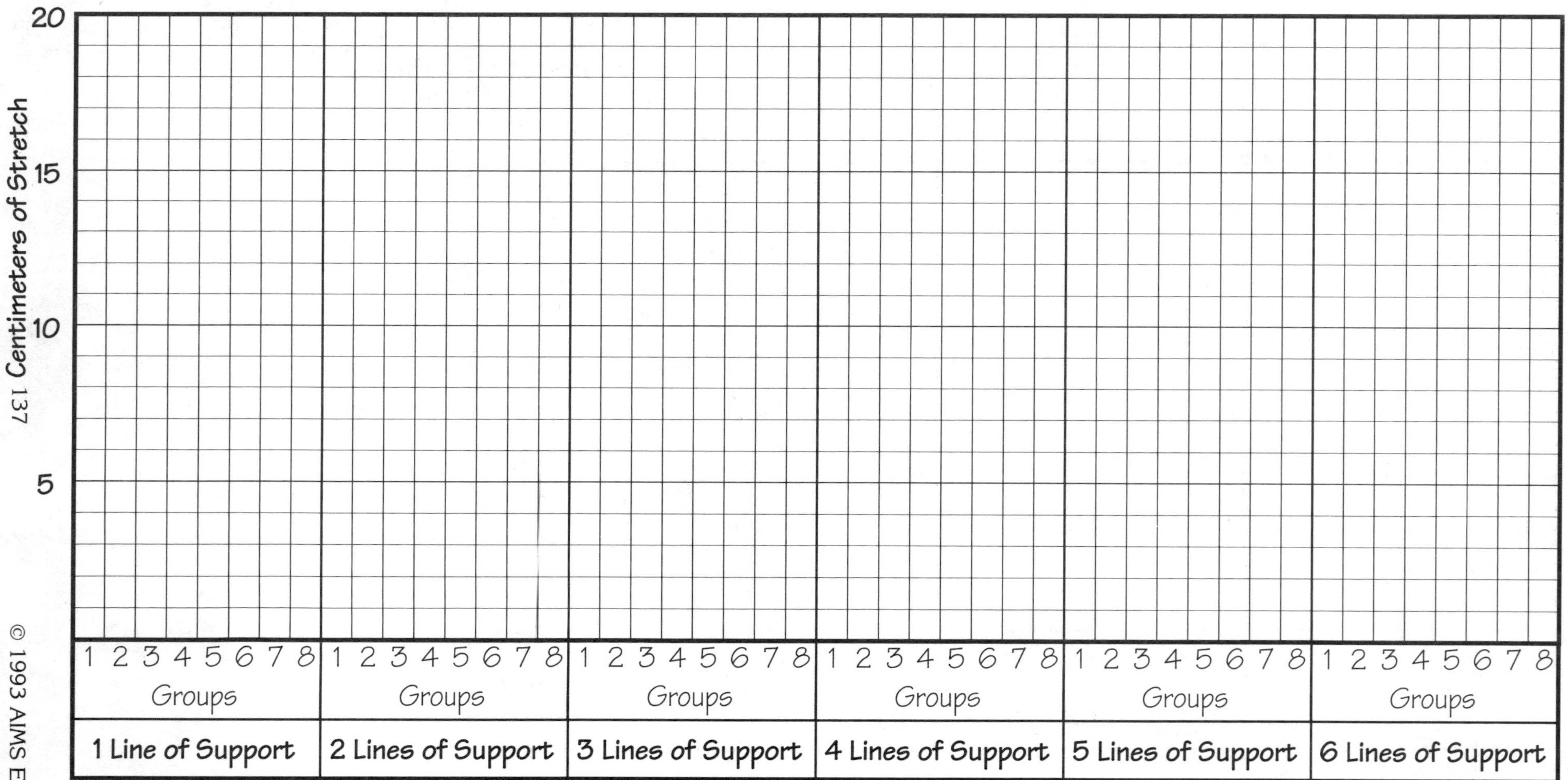

Explain what generalizations you can make from information shown on the graph.

If you kept adding more pulleys and their lines of support, would the centimeters of stretch ever reach zero? Why?

June 20

Dear Journal,

Mucky and I were playing some basketball at his house this afternoon. We got in an argument when Mucky accused me of cheating by moving too fast. He said he did not care because he knew he was stronger than me anyway. I told him there was no way! We decided to have an arm wrestling contest to decide who was stronger. Guess we should have realized how hot it was before we got ready to have our match in the sunny front yard.

We were arm wrestling with all our might. I was really surprised at how strong Mucky was. His face was turning bright red and dripping with sweat. I was starting to win when Mucky suddenly quit. He was faint, felt sick, and began getting pale. I knew he was suffering from heat exhaustion and that I had to get him inside the cool house, but he was too weary to move. I knew I had to act quickly before he got worse. I couldn't lift Mucky's 300 kilograms up the half meter of steps to the porch. I had to think of a way to get him up. I thought of all the machines in the shop, then I remembered the box labeled "Inclined Planes." There was nothing inside but different ramps. That was just the idea I needed–I could roll Mucky up a ramp instead of using the steps. I raced to Mucky's garage and found a long board and made a ramp of it from the sidewalk to the porch. I rolled Mucky up the ramp and into the house.

Mucky was fine after he cooled down, drank some water, and rested. I guess we learned we should not work so hard when it is so hot outside.

June 21

Dear Journal,

Because of the hot weather and not wanting to repeat yesterday's episode, Mucky and I chose a more relaxing pastime on this first day of summer. We were in the machine shop discovering the secrets of one of the last boxes–the one that gave me the idea for how to rescue Mucky yesterday–"Inclined Planes."

I decided there must be more to this box than I thought. But when I looked inside there was just a bunch of ramps. It was then that I remembered the secret of all machines. **If a machine multiplies force, it does so at the expense of distance.**

I decided to do some math myself and figure out how I had gotten Mucky up that inclined plane yesterday.

I remembered Handy telling me that machines just transfer work. The work that is put into the machine must come out of the machine. With Mucky, the work I had to get out was to lift 300 kilograms 0.5 meters. I figured I could at least push with a force of 50 kilograms. So I pulled out this paper and did my calculations.

Work Out		Work In
0.5 m X 300 kg	=	______ m X 50 kg
150 units of work	=	______ m X 50 kg

Tomorrow I am going over to Mucky's house to measure the board and see if it works out. I figure that my small force was multiplied by the inclined plane to be big enough to move Mucky's huge weight. The machine would have to multiply the force at the expense of distance. So the board must be a lot longer than the half meter I would have had to lift Mucky if I had lifted him straight up.

Editor's Note: Fantastic Force had to measure the board to determine how long it was. Try to figure out how long the board had to be from Fantastic's figures?

June 22

Dear Journal,

I was so excited that I had figured out a machine all by myself that when we saw Handy hoeing in his garden, Mucky and I went to tell him about the events of the past couple of days. He was glad to hear that I was really beginning to understand all about work and machines. He did say that arm wrestling was not a very good comparison of abilities because it only compares the strength of two people, and to be fair you should compare power. He told us that Mucky is obviously stronger than I because he has the force to lift his body straight up the steps when he needs to. He told me I had such little force that I had to lift Mucky's weight up slowly along a long ramp. Mucky is stronger than me because he can lift more weight than I can at one time.

Handy suggested we see who is more powerful. He got a pitch fork and a wheelbarrow and told us to follow him. Mucky and I followed Handy over to his compost pile. Handy told us that we were each to take a turn and see who could be the fastest at filling the wheelbarrow with compost. Mucky went first. When Handy said to start, Mucky sluggishly began to lift huge loads of compost into the wheelbarrow. It took him only five lifts to complete the job, but he did it so slowly that it took him three minutes to do it. Handy thanked Mucky for filling the wheelbarrow and took it to his garden to use as mulch. For my turn, I found I could lift only small loads, but I could lift them much faster than Mucky. I must have made 30 or 40 lifts in the two minutes and 50 seconds it took to fill the wheelbarrow. Handy declared me the most powerful as he hauled this load of compost to the garden. Mucky and I did the same amount of work, we lifted compost to fill a wheelbarrow. I was more powerful because I could do that amount of work faster.

Handy thanked us for helping him mulch his garden and then suggested we look up power in The Mechanical Encyclopedia. Under the word "power" it said:

Power is the amount of work done divided by the time it takes to get that work done. Power is calculated with the formula

$$\text{Power} = \frac{\text{Work}}{\text{Time}} = \frac{\text{force} \times \text{distance}}{\text{Time}}$$

The number that you get for power means you could compare people doing different amounts of work. This means I can compare myself to all my friends. Maybe we should have a power contest this summer.

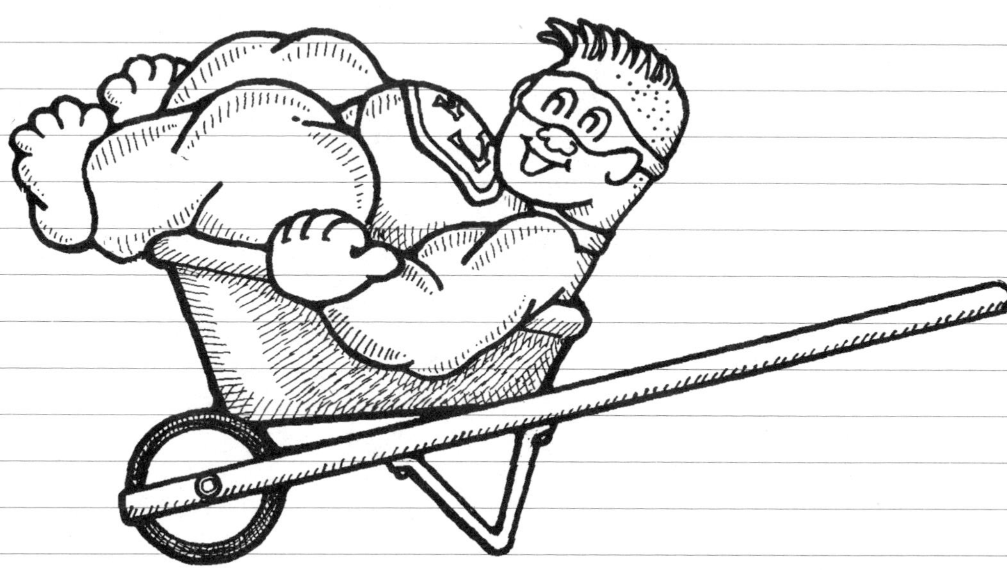

Making the Grade

Topic Area
Inclined planes

Introductory Statement
In this investigation, students will find that it takes less force to lift (push or pull) an object along an inclined plane than it does to lift it straight up. The amount of work done will remain the same, but the force exerted will be less.

Math
Measuring
 length
Graphing
Using whole number operations
Using calculators

Science
Physical science
 Simple machines

Math/Science Process Skills
Observing
Comparing and contrasting
Collecting and recording data
Making and testing hypotheses
Collecting and recording data
Interpreting data
Applying and generalizing

Materials
Per group:
 sturdy wooden meter stick
 rubber band (6–8 cm in length)
 2 large paper clips
 1/2 lb. margarine tub
 60 cm string
 masses–400 grams (see *Management 5*)
 stack of books (20 cm high)
 colored pencils
 rulers–30 cm
 calculators
 bathroom scale

Key Question
How much force does it take to lift 400 grams 20 cm in height?

Background Information
An *inclined plane* is a simple machine that reduces the force needed to lift an object. By moving an object up an inclined plane, less force is used to get the object up to a certain level than if the object were lifted directly to that same level. The longer the plane, the more gradual the slope, the less force needed to move the object. The shorter the plane, the steeper the slope, the more force necessary to move the object. The object, however, must now be moved a longer distance as it travels up the inclined plane.

Ideally, the work done to lift an object straight up is the same as the work done to move the object up an inclined plane. With the inclined plane, the force is spread over a longer distance. However, friction is always a part of our world and, as a result, the outcome is never quite the ideal. The greater the amount of friction,the greater the outcome is from the calculated ideal.

Work is found by multiplying force times distance. To make this activity useable to classrooms without spring scales, it uses the non-standard unit of rubber band units (rbu) for force. Rubber band units are actually centimeters of the stretch of a rubber band. However, students may become confused if centimeters were used as both distance and force measures. By using rbus, students should be able to differentiate between the force units and the distance units.

Management
1. Allow two, 45-minute periods for this activity.
2. This lesson may be done as a whole class activity if the class is not ready to work in independent groups. If there is room and materials, groups of five students work well. Each group will need:
   - a Holder–one who keeps the plane steady;
   - a Measurer–one who observes and states the amount the rubber band stretches;
   - a Lifter/Puller–one who holds the ruler and pulls the paper clip at the same time;
   - a Recorder–one who records the amount of stretch from the rubber band;
   - an Observer–one who keeps check on all the activities and lets others know if things are going awry.

   These roles should be designated and responsibilities covered before proceding with the actual activity.
3. The teacher should try the investigation before doing it with the class. This will allow the teacher to see how to measure the centimeters of stretch of the rubber band. The measurement of the rubber band **must be taken** while it is in motion, otherwise there is a tendency to let it relax which will result in greater inaccuracy. This method of measurement may need to be demonstrated a few times. Students need to see how to hold the paper clip and ruler at the same time they are pulling and measuring. The students should practice this before they proceed.
4. The amount of stretch in the rubber band (rbu) should be measured to the nearest centimeter.
5. Prior to the activity, weigh the containers. Add masses to the containers until each weighs about 400 grams. This weight will provide an adequate stretch for most rubber bands. Experimentation will enable you to determine whether a rubber band of another size needs

to be used. It is important that the weight remains the same throughout the activity.

6. The line of string pulling the container must be parallel to the plane.
7. Emphasize that the final height of the plane remains constant at 20 centimeters; only the length differs.
8. As students begin to pull the containers, they may have to adjust the strings so that the container does not slide off the plane.
9. Sturdier planes may be made by placing two meter sticks side by side and taping them together on their underneath sides or by using strips of plywood or heavy cardboard instead of the meter sticks. When selecting alternatives for the inclined planes, keep in mind the amount of friction that will be produced with different surfaces.
10. The teacher may have an inclined plane made to be used for demonstrating, but should not do any measuring of stretch.

Procedure

1. Discuss the *Key Question* with the students.
2. Review the vocabulary: resistance–a force, which in this case, is measured by the amount of net stretch in the rubber band (rbu); effort; work; height; and distance.
3. Using the demonstration model of the inclined plane, show students examples of the rubber band while it is at rest and the rubber band while it is stretched. Explain that the difference in the measurements of the rubber band in these two settings is the net stretch which will be measured in centimeters but recorded as rubber band units (rbu).
4. Distribute materials and the introductory handout. Have the students measure the rubber band at rest, construct the container as shown in the instructions, and predict the net stretch of the rubber band.
5. Distribute the record sheet.
6. Lifting the container straight up, have students measure the stretch of the rubber bands, recording the lengths and completing the *Straight Lift* row of the chart.
7. Instruct students to complete the chart as they do the rest of the activity.
   - The distance of the pull is the length of the inclined plane. Students should adjust the length of the meter stick until the 30 cm mark is on the edge of the books. (See diagram)
   - Have the students predict the amount of stretch that will occur in their rubber bands when they pull the container up the inclined plane. (Predictions should be made for each trial and then tested before going on to the next distance of pull. This strategy not only increases active participation, but allows students to assess the data and form better predictions.
   - Direct the students to pull the container up the inclined plane, measuring and recording the length of the stretched rubber band in rbus.
   - Have students adjust the length of the plane to the next longer length and repeat the same procedure, continuing this sequence until the 100 cm length is reached.

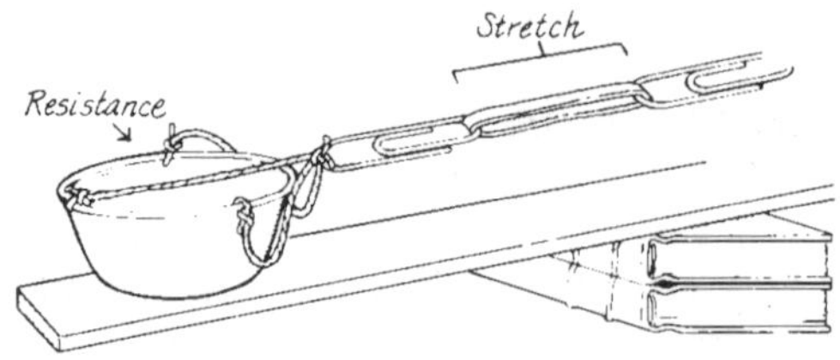

8. Have students examine their results. Discuss why the numbers may not be exact. [measured numbers are at best approximations, never exact]
9. Distribute the graph. Review the variable labels with the students. Have them then complete the graph using their data from the chart. Be sure the students graph the straight lift as one piece of data. The straight lift has no slope and is the most efficient way to lift. You must, however, exert more force to lift the object, but for a shorter distance.
10. The students should recognize that the 20 cm straight lift, and the 20 cm that the 100 cm inclined plane is elevated, raises the container the same amount. The only difference is that the tub traveling on the 100 cm inclined plane must travel 100 cm to be raised the 20 cm. They should also notice that the force is reduced as the distance increases.
11. At the conclusion of this activity, the last activity page may be used as assessment or reinforcement.
12. Students should determine the force they can push by placing a bathroom scale against a wall and push against it with all their might. There should be a designated recorder and observer to determine the amount of force each student exerts.
13. Students use their data to determine the length of inclined plane they would each need to push Mucky Mass into the house. Students may use the chart to calculate their answer. This will be an ideal answer and will not take into account friction.

Discussion

1. Have the groups compare their graphs to see what similarities they have. [The general pattern that the effort force gets smaller as the length of the inclined plane increases should be evident in all the graphs.]
2. What generalization about the inclined plane can be made from the graphs? [The effort force gets smaller as the length of the inclined plane increases.]
3. How could you use an inclined plane to help you?
4. Why did the force go down as the length of the plane got longer? [The slope decreased] Did you reach the same height? [Yes, 20 cm, but it took a longer distance to reach the height.]
5. When you lift the tub in the direct lift, what happens to the rubber band at the very beginning of the lift? [The rubber band jerks] What might cause this? [overcom-

ing inertia takes more energy than it does to keep an object moving once it's started]

6. What factors could make a difference in the amount of force needed to pull the tub up the inclined plane? [friction, the smoothness of the container, the roughness of the plane, the weight to be pulled, the shape of the container]
7. What could be done to minimize the friction? [wax the tub and/or the inclined plane, use an object with wheels or rollers]
8. Theoretically, comparing the amount of force times the distance of pull for each example should give you the same results. In the activity you will find that these are not exactly the same. What causes the difference? [friction, efficiency of the plane, inaccurate measurement]

Extension

Do the AIMS activity *Simply Marbleous* in *Math + Science–A Solution.*

Major Conceptual Components

- Energy is used to accomplish work.
- Work is done when something is moved.
- The amount of work done is determine by how much force was used to move the object and how far the object was moved.
- Inclined planes are flat, sloping surfaces over which objects can be pushed or pulled to move them to a higher level.
- Inclined planes can be used to raise objects with less force than a straight lift.

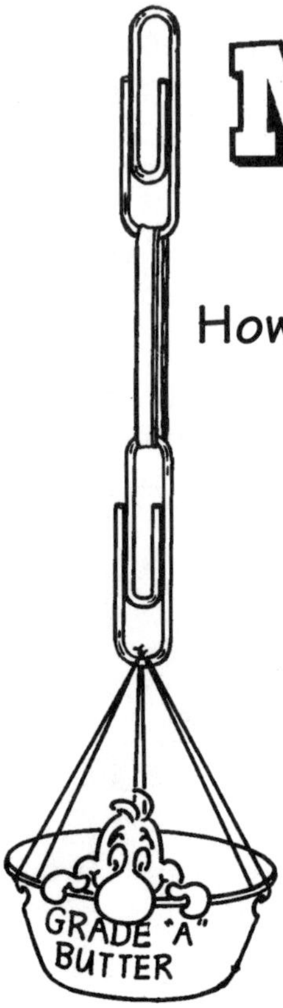

MAKING THE GRADE

How much force does it take to lift 400 grams 20 cm in height?

Construct the Container

A. Carefully poke 3 holes in the plastic tub (hole punch).
B. Thread the string through the holes and tie.
C. Attach a jumbo sized paper clip to the end of the strings.
D. Attach the other end of the clip to the rubber band.
E. Attach another jumbo clip to the other end of the rubber band. Use this clip to pull the tub.
F. Add weights until the tub weighs 400 grams.

1. Measure rubber band at rest. __________ rbu

2. When you lift the 400 gram butter tub, how far will the rubber band stretch?

Total Stretch		Band at Rest		Net Stretch
rbu	-	rbu	=	rbu

3. When you pull the container on the inclined plane (holding the paper clip only), how far will the rubber band stretch?

Total Stretch		Band at Rest		Net Stretch
rbu	-	rbu	=	rbu

4. Did you think the rubber bands would stretch the same amount? Why?

5. Which one (the straight lift or the inclined plane) will use more force?

6. Do the activity to check your guess.

MAKING THE GRADE

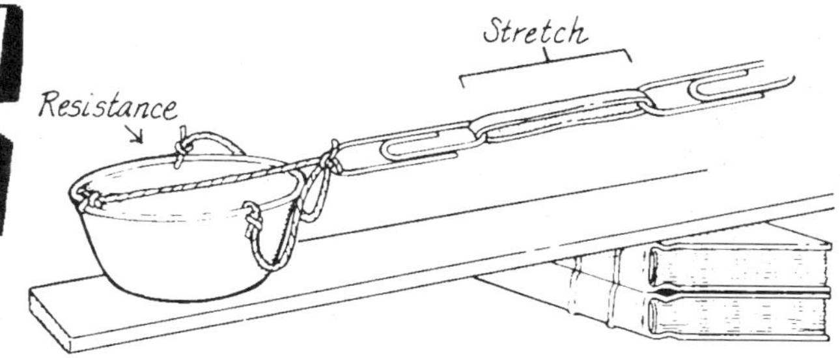

How does the length of an inclined plane affect the force?

	Predicted Net Stretch (rbu)	Actual Stretch (rbu)	−	Band at Rest (rbu)	=	Force Net Stretch (rbu)	X	Distance of Pull (cm)	=	Work (f x d) (rbu - cm)
Straight Lift			−		=		X	20	=	
Inclined Plane			−		=		X	40	=	
			−		=		X	50	=	
			−		=		X	60	=	
			−		=		X	70	=	
			−		=		X	80	=	
			−		=		X	90	=	
			−		=		X	100	=	

What do you notice about the amount of force being used? Write 2 sentences.

What do you notice about the amount of work done in a straight lift and an inclined plane?

Write a formula for how to find the amount of work done.

MAKING THE GRADE

1. Graph the rubber band units of stretch (force) for the distance the container (resistance) traveled. Use your data from the chart.

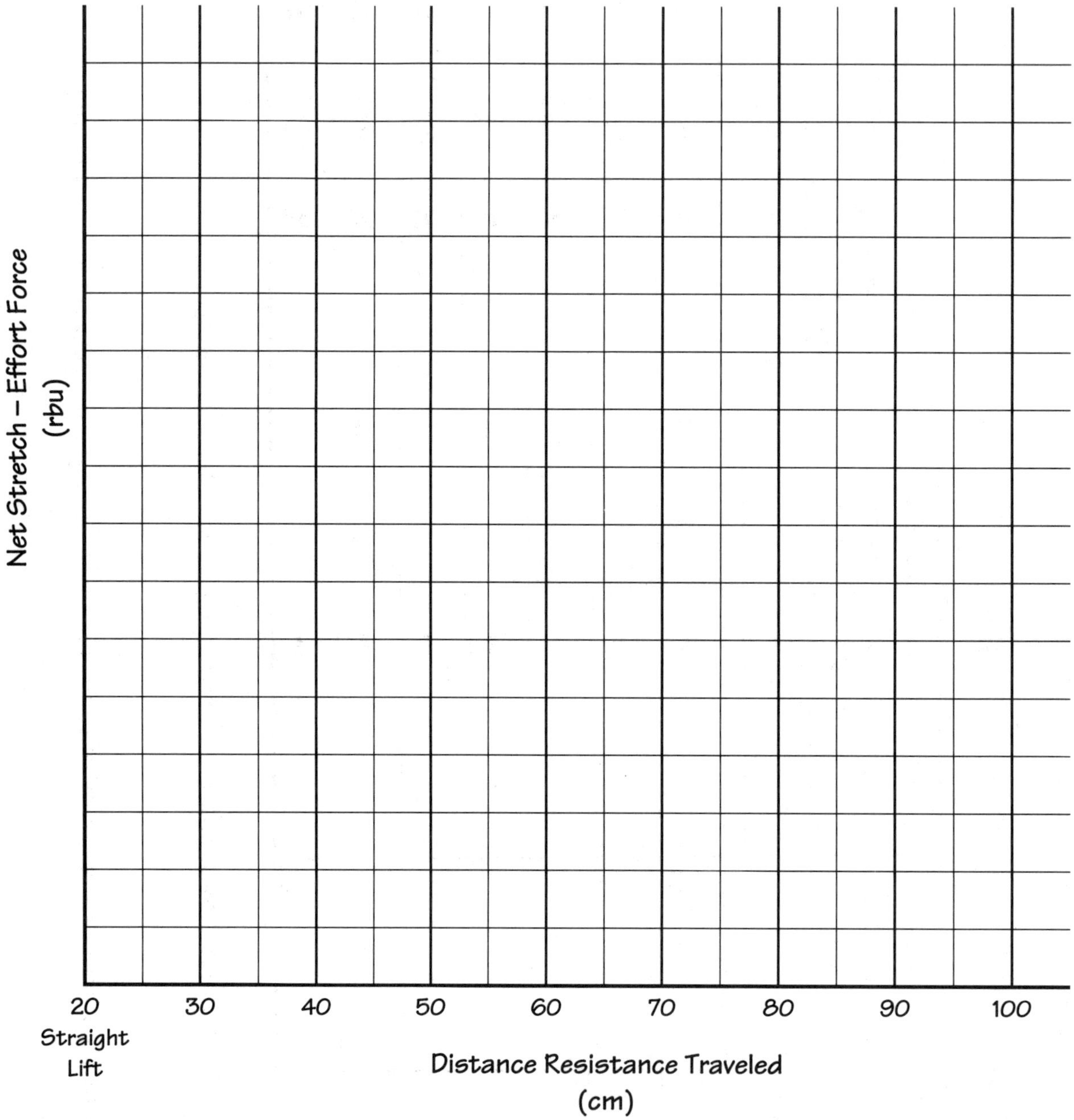

2. Height lifted with straight lift = _____ cm   Height lifted with inclined plane = _____ cm

3. Make a generalization about the use of an inclined plane as compared to a straight lift.

4. List 3 examples of inclined planes used in everyday life.

Mucky MAKES THE GRADE

Mucky Mass has fainted, and you must *get* him inside the house?
How long *does* your inclined plane need to be?

Mucky Mass, who has fainted, weighs 300 kg. The height from the ground to the door is _____ m.

Push against a bathroom scale to determine the amount of pushing force you have. ________kg (my force)

	Work In			Work Out		
Student	Force ×	Distance =	Total Work (f x d)	Mucky's Mass ×	Lift =	Total Work (m x l)
1.						
2.						
3.						
4.						
5.						

How long would the inclined plane have to be if all of the members of your group pushed Mucky?

The Plane Truth

Topic Area
Efficiency of inclined planes

Introductory Statement
The students will make and compare different surfaces of inclined planes, testing the efficiency of each surface.

Math
Finding percentages
Estimating
Graphing
Measuring length
Using rational numbers
percents

Science
Physical science
Simple machines

Math/Science Process Skills
Predicting
Inferring
Observing
Making and testing hypotheses
Collecting and recording data
Applying and generalizing

Material
For each group:
2 wooden meter sticks
1/2 lb. margarine tub
string
ruler (30 cm)
books (stack 20 cm high)
rubber bands (6-8 cm long)
2 large paper clips
colored pencils
masses (300-500 gm)
For the centers:
5 cm x 100 cm strip of tagboard
2 sheets of sandpaper
1 balance
waxed paper
cooking spray or spray lubricant
plastic coated playing cards
1 Hot Wheels®-type car
rubber bands

Key Question
How will the surface of an inclined plane affect its efficiency?

Background Information
An inclined plane is a simple machine. To push or pull an object up an inclined plane takes less force than to lift it straight up, but it must travel a longer distance to achieve the same height.

Efficiency of a machine tells how well the machine transfers energy. It is calculated with the following formula:

$$\frac{\text{Amount of work a machine can produce (output)}}{\text{Amount of work you put into a machine (input)}} \times 100$$

Efficiency is measured by percent and the larger the percent, the more efficient the machine. Extra force is needed to overcome friction. The work output is always less than the work input. This loss of work causes the efficiency of a machine to be less than 100%.

A straight lift is the most efficient method of lifting an object because it is the most direct way to move the mass (takes the shortest distance); it also requires the most force. Friction also interferes least with the straight lift.

In this activity, efficiency can be approximated by dividing the work of the straight lift (output) by the work done using the inclined plane (input) and multiplying by 100 to find the percentage.

$$\frac{\text{Work using straight lift}}{\text{Work using inclined plane}} \times 100$$

Management
1. This activity should follow *Making the Grade*. It is organized to be set up as six centers. Groups of students will begin at an assigned center, complete it, and move on to the next center as per teacher directions.
2. Students should be able to make the different plane surfaces themselves; however, to save class time, the teacher may choose to assemble the inclined planes so they are ready for student use.
3. The class should be divided into six groups with groups rotating through the different centers. They will compare the amount of work done and force used for each of the different surfaces to those of the straight lift.
4. Student groups should be divided into measurer, puller, recorder, and holder. These jobs are the same as outlined in *Making the Grade*.
5. The teacher may want to use a kitchen timer to signal students working at each center when it is time to rotate. At least 10 minutes is needed for the <u>first</u> rotation because the students must make the planes and the containers (students may use the containers from *Making the Grade*). If some groups finish early, they may retest their plane to check accuracy.
6. Measuring the stretch of the rubber band will be the most difficult part. If the puller can manage to hold the paper clip and the ruler at the same time, it is much easier to get an accurate measurement. The measurement must be taken while moving. After doing *Making the Grade*, this task should be easier.

7. Allow two 45-minute periods for this activity.
8. Before the activity, determine six areas for the stations. Place a copy of the instructions and the appropriate materials at each station.

Procedure

1. Discuss the *Key Question* and have students discuss possible ways to test their ideas.
2. Distribute the record sheet and have each group go to a station.
3. Have each group follow the instructions to assemble their container for the straight lift.
4. Demonstrate for the students how to measure the stretch of the rubber band and make sure they understand this procedure.
5. Have students estimate the amount of stretch before they take their measurements for a straight lift. Students record this information on the chart on the record sheet.
6. Have students adjust their containers for the inclined plane. Other students at the same time can assemble the inclined plane assigned to the group. Make sure that each group makes the inclined plane 90 cm long and with a vertical rise of 20 cm.
7. Instruct students to estimate how much the rubber band will stretch when it is used to pull the container up the inclined plane.
8. Students then pull their 400 gram container up the inclined plane using the stretch in the rubber band to measure the force required to move the container. Students record the measurements on the chart. Although the measurements will be in centimeters of stretch, students will record their findings in rubber band units (rbu).
9. Students rotate completing steps seven and eight at every station. All measurements are recorded on the chart.
10. Distribute the graph and have students complete it using the data from their charts. Students are to make a bar graph putting the planes in order of efficiency from left to right.

Discussion

1. Have groups compare their order of efficiency to see if they are in agreement.
2. What is the most efficient way to lift something? [straight lift] Why? [little friction]
3. Which inclined planes appear to be more efficient?
4. What makes some inclined planes more efficient than others? [less friction]

Major Conceptual Components

- Energy is used to accomplish work.
- Work is done when something is moved.
- The amount of work done is determine by how much force was used to move the object and how far the object was moved.
- Friction is a force that resists motion.
- Friction can be reduced.
- Machines lose efficiency because of friction.
- Efficiency is the measure of how effectively a machine has transferred energy.
- Efficiency is the ratio of work output to total input; or the percentage of work put into a machine that is converted to useful work output.
- Inclined planes are flat, sloping surfaces over which objects can be pushed or pulled to move them to a higher level.
- Inclined planes can be used to raise objects with less force than a straight lift.

Container for Straight Lift

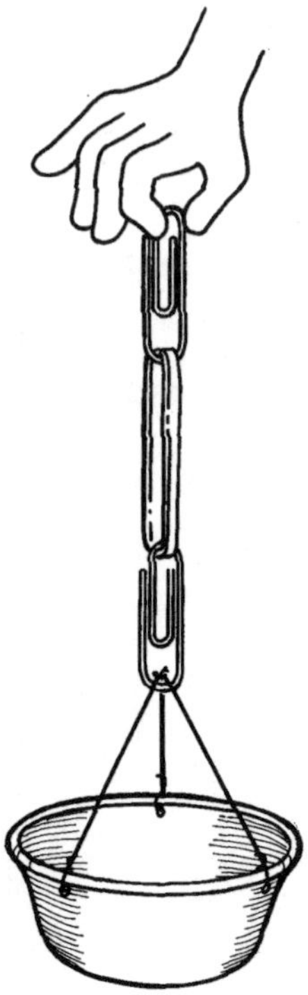

A. Carefully poke 3 holes in the plastic tub (hole punch)
B. Thread the string through the holes and tie.
C. Attach a jumbo-sized paper clip to the end of the strings.
D. Attach the other end of the clip to the rubber band.
E. Attach another jumbo clip to the other end of the rubber band. Use this clip to pull the tub.
F. Add weights until the tub weighs 400 grams.

To adjust the container for the inclined plane, two of the three strings may be cut. If the cut strings get in the way of the sliding container, trim them and put the extra string in the container.

Wooden Inclined Plane

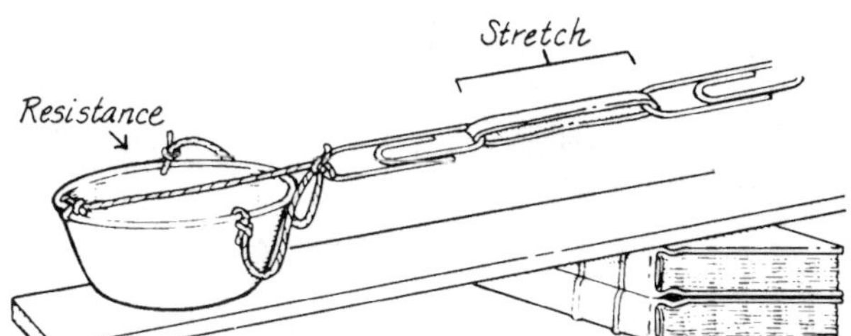

A. Prepare a container (tub) for the inclined plane.
B. Make a stack of books 20 cm high.
C. Tape 2 meter sticks together.
D. Lean the meter sticks against the books and adjust it so the inclined plane is 90 cm long.
E. Tape the bottom of the inclined plane so it does not slide.
F. Place the tub at the bottom of the plane.
G. Pull the tub along the inclined plane and measure the length of stretch in the rubber band.
H. Record your information on the chart.

Sandpaper Inclined Plane

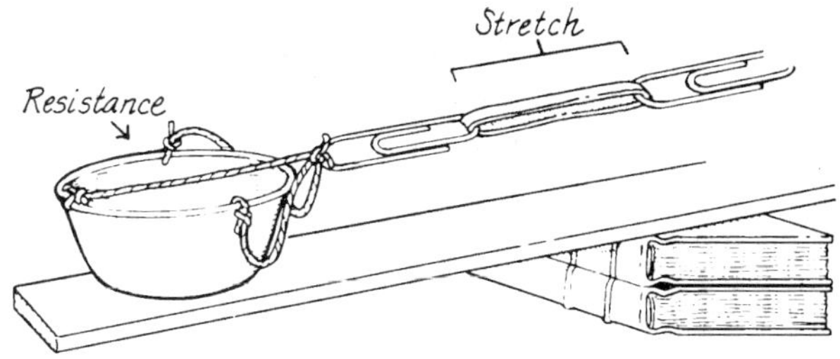

A. Prepare a container (tub) for the inclined plane.
B. Make a stack of books 20 cm high.
C. Tape 2 meter sticks together.
D. Lean the meter sticks against the books and adjust it so the inclined plane is 90 cm long.
E. Tape the bottom of the inclined plane so it does not slide.
F. Cut sandpaper into strips as wide as the 2 meter sticks.
G. Tape the strips of sandpaper down on top of the meter sticks so the sandy part faces up.
H. Place the tub at the bottom of the plane.
I. Pull the tub along the inclined plane and measure the length of stretch in the rubber band.
J. Record your information on the chart.

Plastic Cards Inclined Plane

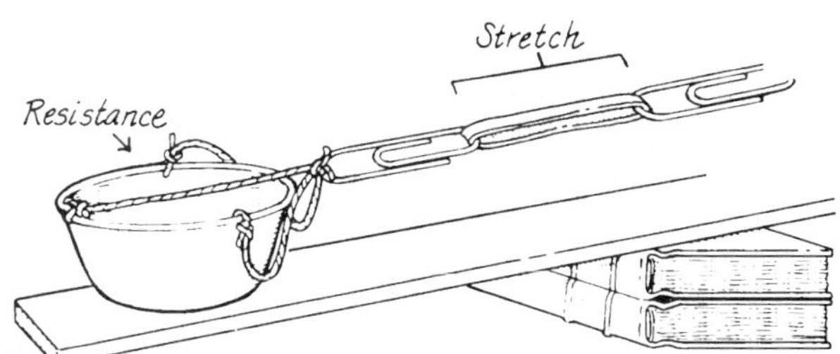

A. Prepare a container (tub) for the inclined plane.
B. Make a stack of books 20 cm high.
C. Tape 2 meter sticks together.
D. Lean the meter sticks against the books and adjust it so the inclined plane is 90 cm long.
E. Tape the bottom of the inclined plane so it does not slide.
F. Place plastic playing cards in a row on top of the meter sticks.
G. Tape the cards down on top of the meter sticks. Make sure the edges line up perfectly.
H. Place the tub at the bottom of the plane.
I. Pull the tub along the inclined plane and measure the length of stretch in the rubber band.
J. Record your information on the chart.

Wax Paper Inclined Plane

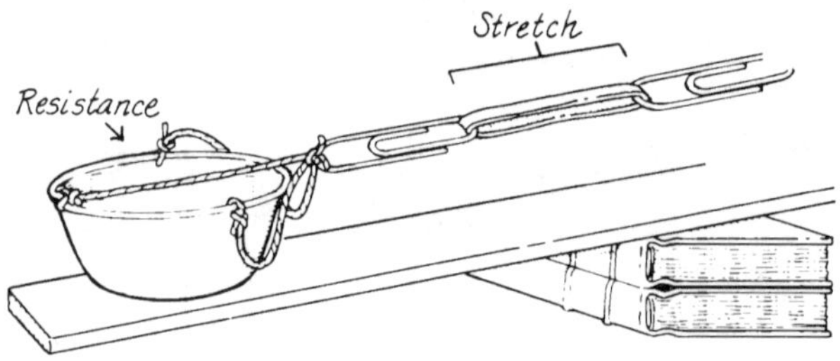

A. Prepare a container (tub) for the inclined plane.
B. Make a stack of books 20 cm high.
C. Tape 2 meter sticks together.
D. Lean the meter sticks against the books and adjust it so the inclined plane is 90 cm long.
E. Tape the bottom of the inclined plane so it does not slide.
F. Cut the wax paper into strips as wide as the 2 meter sticks.
G. Tape the strips of wax paper down on top of the meter sticks.
H. Place the tub at the bottom of the plane.
I. Pull the tub along the inclined plane and measure the length of stretch in the rubber band.
J. Record your information on the chart.

Wax Paper and Spray on Inclined Plane

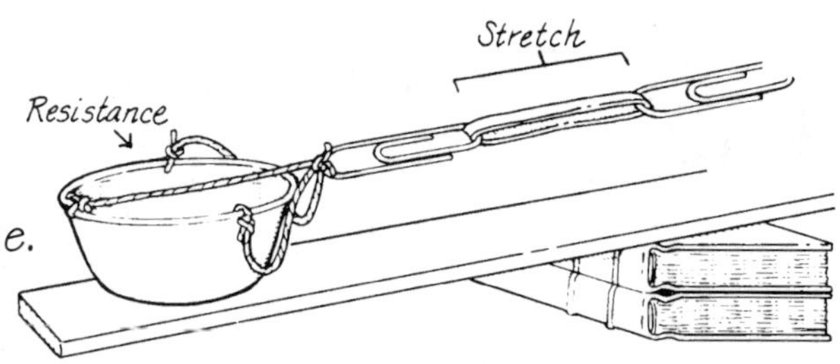

A. Prepare a container (tub) for the inclined plane.
B. Make a stack of books 20 cm high.
C. Tape 2 meter sticks together.
D. Lean the meter sticks against the books and adjust it so the inclined plane is 90 cm long.
E. Tape the bottom of the inclined plane so it does not slide.
F. Cut wax paper into strips as wide as the 2 meter sticks.
G. Tape the strips of wax paper down on top of the meter sticks.
H. Spray the top of the wax paper and the bottom of the tub with the spray.
I. Place the tub at the bottom of the plane.
J. Pull the tub along the inclined plane and measure the length of stretch in the rubber band.
K. Record your information on the chart.
L. **Wipe off the bottom of your tub** before going to the next station.

Wheels on an Inclined Plane

A. Prepare a container (tub) for the inclined plane.
B. Make a stack of books 20 cm high.
C. Tape 2 meter sticks together.
D. Lean the meter sticks against the books and adjust it so the inclined plane is 90 cm long.
E. Tape the bottom of the inclined plane so it does not slide.
F. Use rubber bands to attach the small car to the bottom of the tub.
G. Adjust the car and the tub so the wheels move easily when the tub is pulled.
H. Place the tub on the wheels at the bottom of the plane.
I. Pull the tub along the inclined plane and measure the length of stretch in the rubber band.
J. Record your information on the chart.

THE PLANE TRUTH

What is the easiest way to get something up an inclined plane?

Predict: Put in order the straight lift and the various inclined planes from most efficient to least efficient.

1. ____________________ 3. ____________________ 6. ____________________

2. ____________________ 4. ____________________ 7. ____________________

5. ____________________

Do the activity and complete the chart to determine which inclined plane is most efficient.

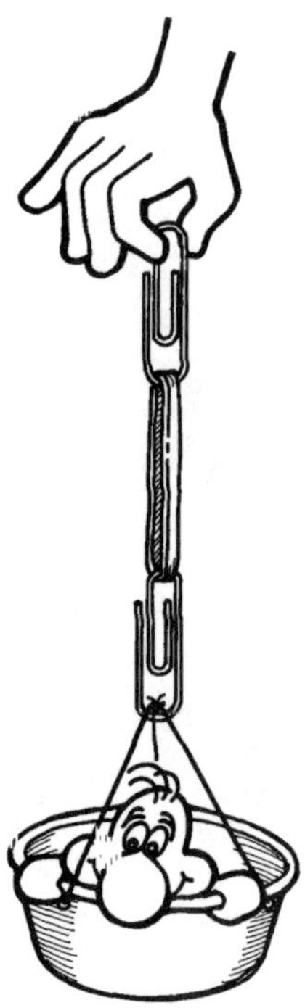

	Estimation of Stretch (rbu)	Actual Stretch (rbu) −	Rubber band at Rest (rbu) =	Net Stretch (rbu) ×	Distance of Lift (cm) =	Work A (rbu • cm)	
Straight Lift							
Inclined Plane	Estimation of Stretch (rbu)	Actual Stretch (rbu) −	Rubber band at Rest (rbu) =	Net Stretch (rbu) ×	Distance of Pull (cm) =	Work B (rbu • cm)	Efficiency (A/B x 100)
Wooden							
Sandpaper							
Plastic							
Wax Paper							
Sprayed Wax Paper							
Wheels							

Conclusion: From the activity, put in order the straight lift and the various inclined planes frommost efficient to least efficient.

1. ____________________ 3. ____________________ 6. ____________________

2. ____________________ 4. ____________________ 7. ____________________

5. ____________________

What do you think causes the different surfaces to be in this order?

Wedge-Ease

Topic Area
Wedges

Introductory Statement
Students will construct wedges from cardboard using the patterns provided. They will then use weights to determine which wedge pushes books apart with the least force. By making a graph, students will discover that the force needed to push through the books is directly related to the length of the wedge.

Math Skills
Measuring
- weight
- length

Estimating
Graphing
Using geometry and spatial sense
Using calculators

Science
Physical science
- Simple machines

Math/Science Process Skills
Predicting
Observing
Collecting and recording data
Interpreting data
Applying
Generalizing

Materials
Per group:
- 3 pieces of cardboard precut to 9" X 12"
- extra pieces of cardboard (see *Procedure 2)*
- tape
- scissors
- utility or craft knife
- 2 books each about 5 cm in thickness
- 300-500 grams of weight

Key Question
Draw three isosceles triangles of different lengths on the chalkboard then ask, "If these three triangles were cross-sections of knife blades, which would make the best knife?"

Background Information
The wedge is an example of two back-to-back inclined planes. Examples of wedges include knives, axes, chisels, or anything with a sharp edge used for cutting or splitting.

The ideal mechanical advantage of a wedge is determined by dividing the length of one of the inclined planes by the thickness at the wedge's widest point.

In this activity, students will construct three wedges each having the same thickness but inclined planes of differing lengths. Theoretically, the wedge with the longest inclined plane will use the least force to split something. The wedges in order from longest to shortest length are wedge B, wedge C, and wedge A. Students should discover that this sequencing of length from longest to shortest is the same as the order of mechanical advantage for the wedges. The sequence is also the inverse of the amount of force required to split or cut something.

Management
1. Precede this investigation with the other activities found in this book dealing with inclined planes.
2. Before this investigation is started, cut the cardboard into 9" X 12" sheets.
3. If you do not have enough weights, it is recommended that different books be weighed with a spring scale. Record their weights for future use in this investigation.
4. Two books, each 5 cm in thickness, will be needed for each group. Dictionaries or text books work well. The wedge will be forced between the books. To help prevent the wedges from sticking, as they pass through the books, place the bound edges of the book together.
5. Encourage the students to gently hold the wedge in a vertical position between the book stacks. The wedge will have a tendency to fall either to the right or left. To eliminate this problem, hold the wedge in a vertical position until it has started on its path.
6. This investigation will require three 50-minute periods: the first for construction, the second for the investigation, and the third for gathering and analyzing class data.
7. Groups of three are recommended for this activity so each member can construct a wedge and be involved in the actual testing.

Procedure
1. Discuss the *Key Question* with the students.
2. Distribute the necessary materials and have the students construct all wedge patterns as shown on the worksheets. If students are using craft knives to cut the wedge shapes, make certain they pad their tabletops with extra cardboard or several layers of newspaper to prevent scarring the tabletops. Encourage students to make a series of shallow cuts rather than trying to cut through the cardboard on the first try. The students assemble the parts of the wedges. Be certain they appropriately label each wedge as A, B, or C.
3. Each student group places two desks next to each other leaving a space of about 6 cm between the edges of the two desks. They will then place a book which is 5 cm in thickness on each of the desks. Slide the books together until they are side by side, with the bindings against each other. Have the students align the bindings

of the books so they are above the open space between the two desks.

4. The students place one wedge directly over the place where both books meet. One student maintains the wedge in the vertical position. Another student will apply weights until the wedge is forced between the books with the top of the wedge even with the top the of the books. You may need to place a piece of cardboard on top of the wedge to serve as a platform to hold the weights. The third student helps with the stacking of the masses and records the results on the chart. This procedure is done three times for each wedge.
5. Students record the results on the worksheet, find the average weight needed for each wedge, and make a bar graph of the information.
6. At the end of the investigation, have students discuss their findings. They should notice that the longer the wedge the less force was required to split the books.
7. Distribute the second activity sheet. The chart is completed as a class activity. The students rank the wedges by the amount of force used to push through the books. Groups then share their average data with the class. Students calculate the class average weight used to push each wedge and measure and record the length of each wedge. Each student completes the line graph and uses a line of best fit to make predictions.

Discussion

1. Which of the three wedges required the greatest amount of weight to spread the books apart?. . .the least amount? [Wedge C required the most mass, Wedge B the least]
2. What relationship is there between the size of the wedge and the effort required to push it through something? [The longer the wedge, the easier it is to push through something.]
3. How is this relationship shown on the graph? [The bar graph shows the longest wedge, B, has the shortest column. On the line graph, the line indicates an inverse relationship with the weight going down as the length gets longer.]
4. What similarities do you see between a wedge and an inclined plane? [similar shape, longer length, easier to move things]
5. What are some of the ways wedges are used? [razor blades, scissors, knives, mauls]
6. What dimensions would you use to construct a wedge that requires the least amount of force to split something? [longest reasonable edge in cross-section]
7. If a wedge were 5 cm in length, how much force would it take to push through the books? How did you determine this?
8. If you knew the force required to split the books was 220 grams, what must the length of the wedge be?

Extension

Have students build wedges of different lengths and predict how much weight it will take to make their wedges spread the books to verify their conclusions. Have the students test their predictions by using the wedges to spread the books.

Major Conceptual Components

- Energy is used to accomplish work.
- Work is done when something is moved.
- The amount of work done is determined by how much force was used to move the object and how far the object was moved.
- Energy comes in many forms.
- Mechanical energy is the energy due to the position or the movement of something.
- A machine is a device for multiplying forces or changing the direction of forces.
- A wedge is two inclined planes placed back to back.

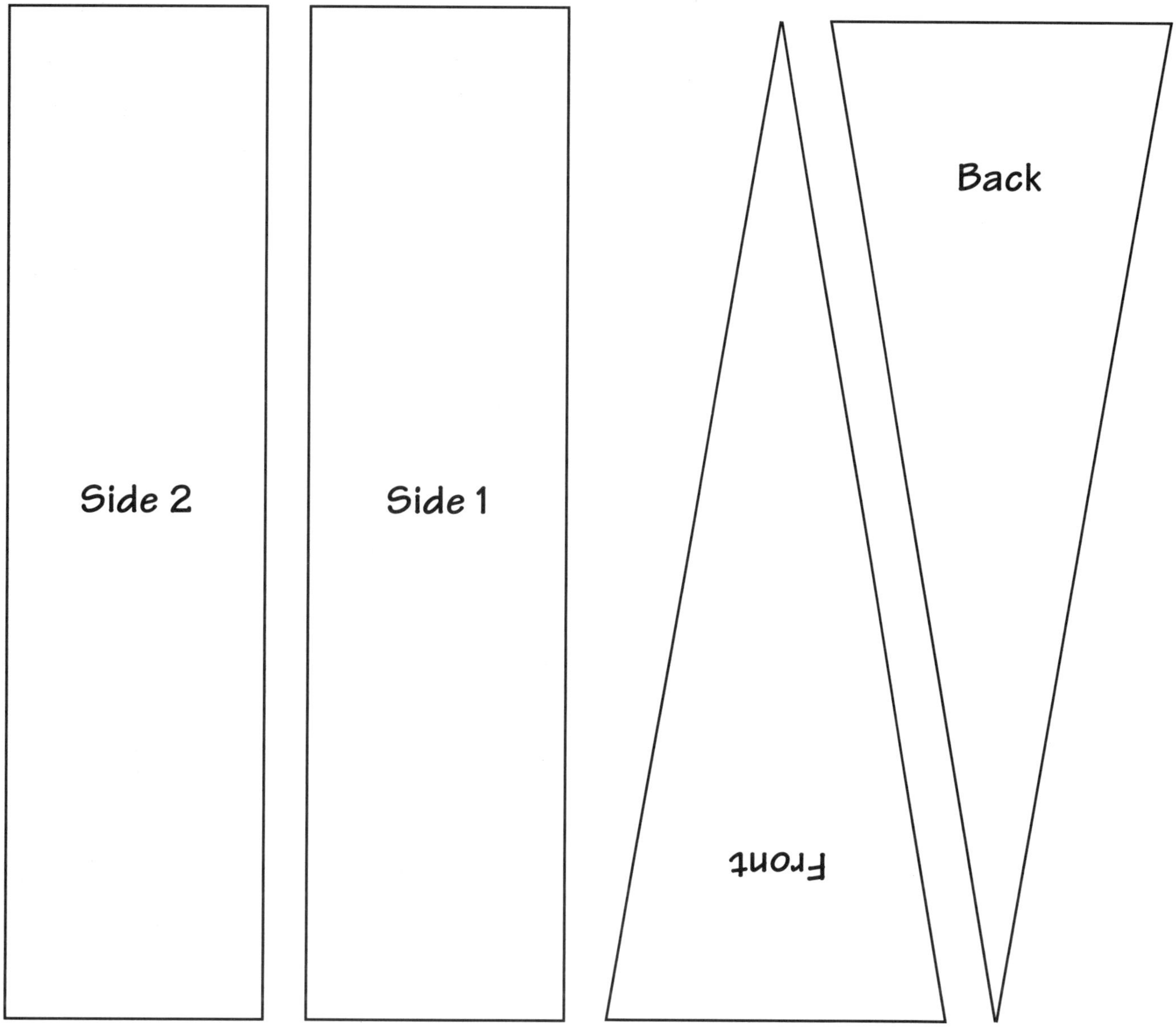

WEDGE-EASE

Wedge A

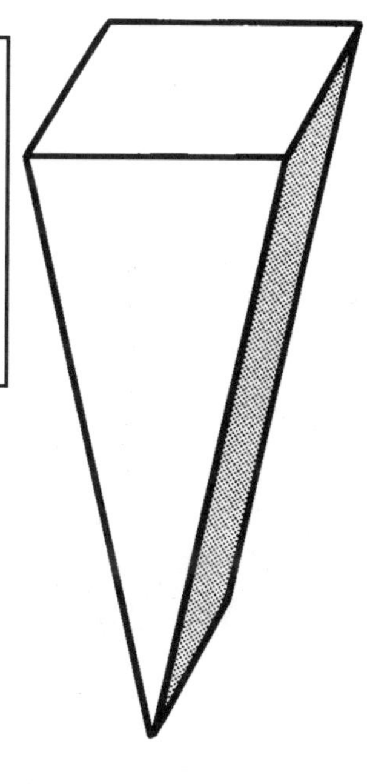

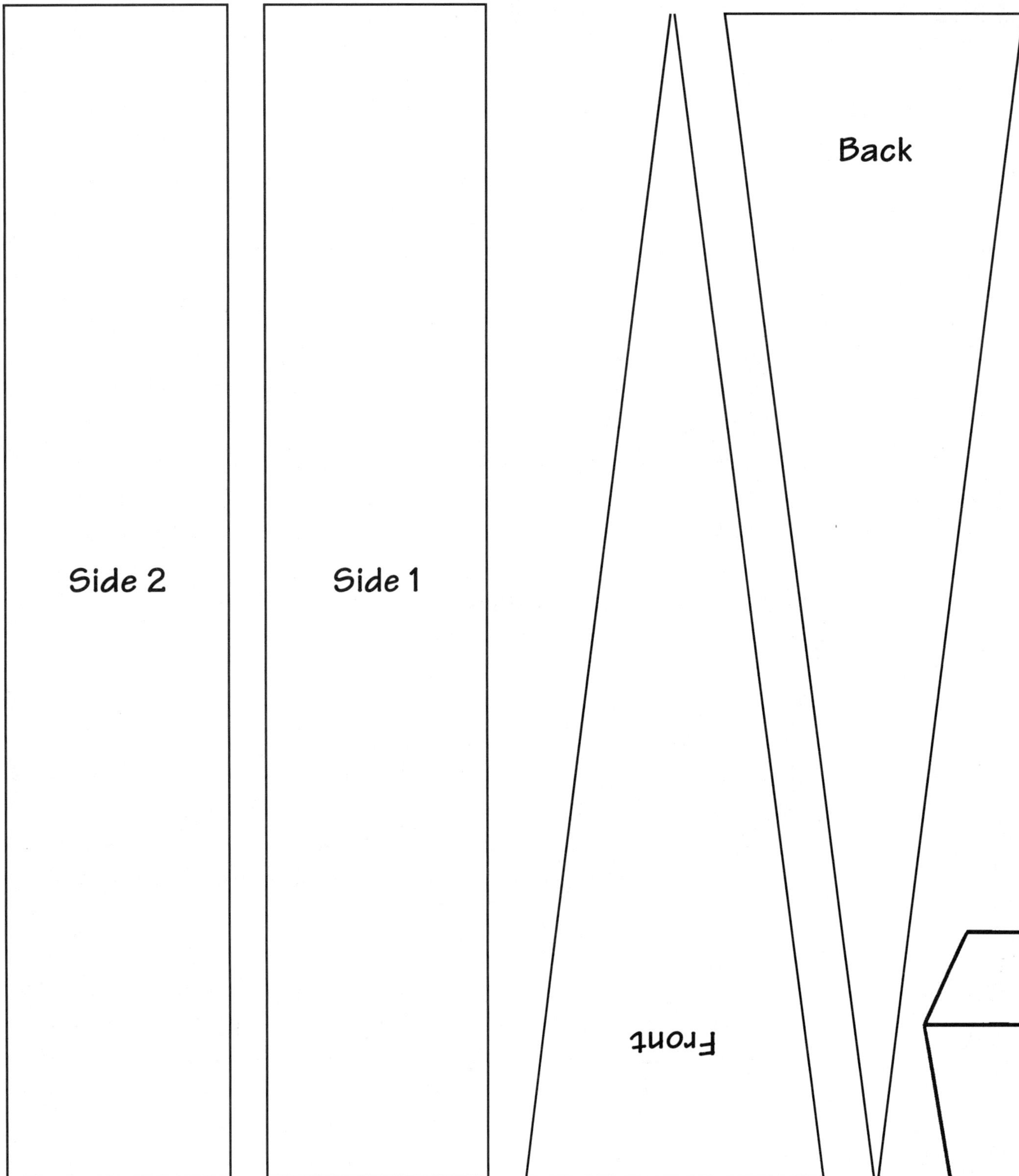

WEDGE-EASE

Wedge B

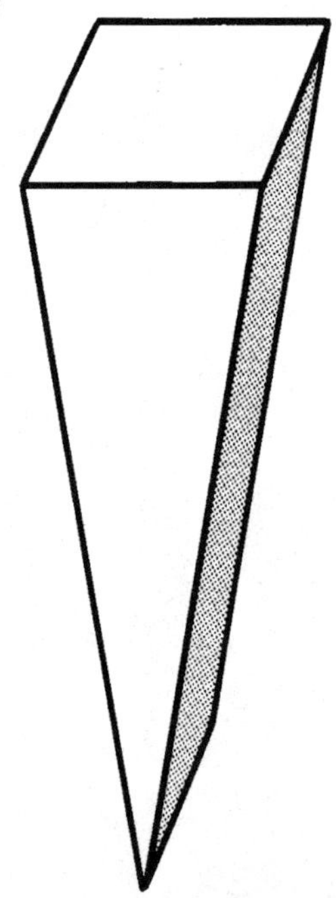

WEDGE-EASE

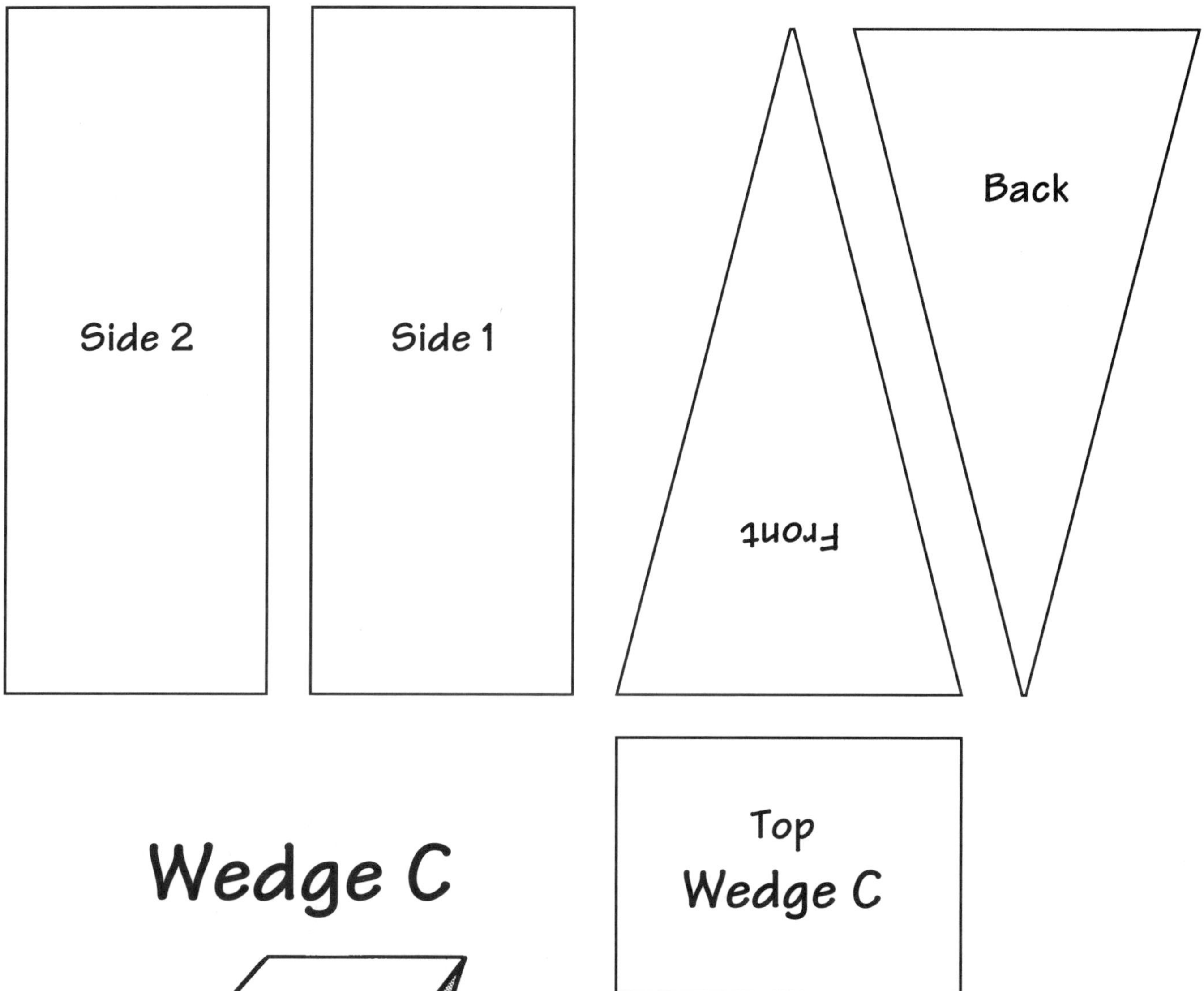

Wedge C

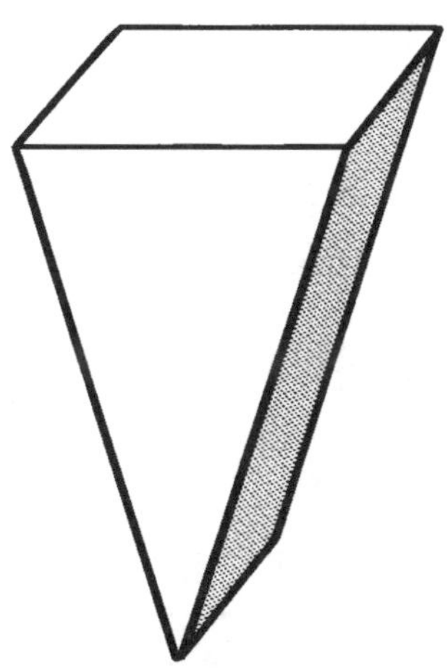

WEDGE-EASE

	Wedge A	Wedge B	Wedge C
Trial 1			
Trial 2			
Trial 3			
Total			
Average			

1. Construct all 3 wedges by pasting the patterns to cardboard, cutting them out, and taping pieces together.
2. Place 2 desks beside each other with about 6 cm of space between them.
3. Place one book on one desk and the other book on the second desk. Each book should have a thickness of about 5 cm. Slide the books together until the bindings are touching and are centered over the space between the desks.
4. Lightly hold a wedge vertically above the place where the books touch. Add weights until the wedge is forced between the books and until the top of the wedge is even with the top of the books.
5. Record how much weight was placed on the wedge to get it to force the books apart.
6. Follow steps 4 and 5 three times for each wedge.
7. Make a bar graph of the data, using a bar for each trial. Draw a line across all three trials to indicate the average weight used for each wedge.

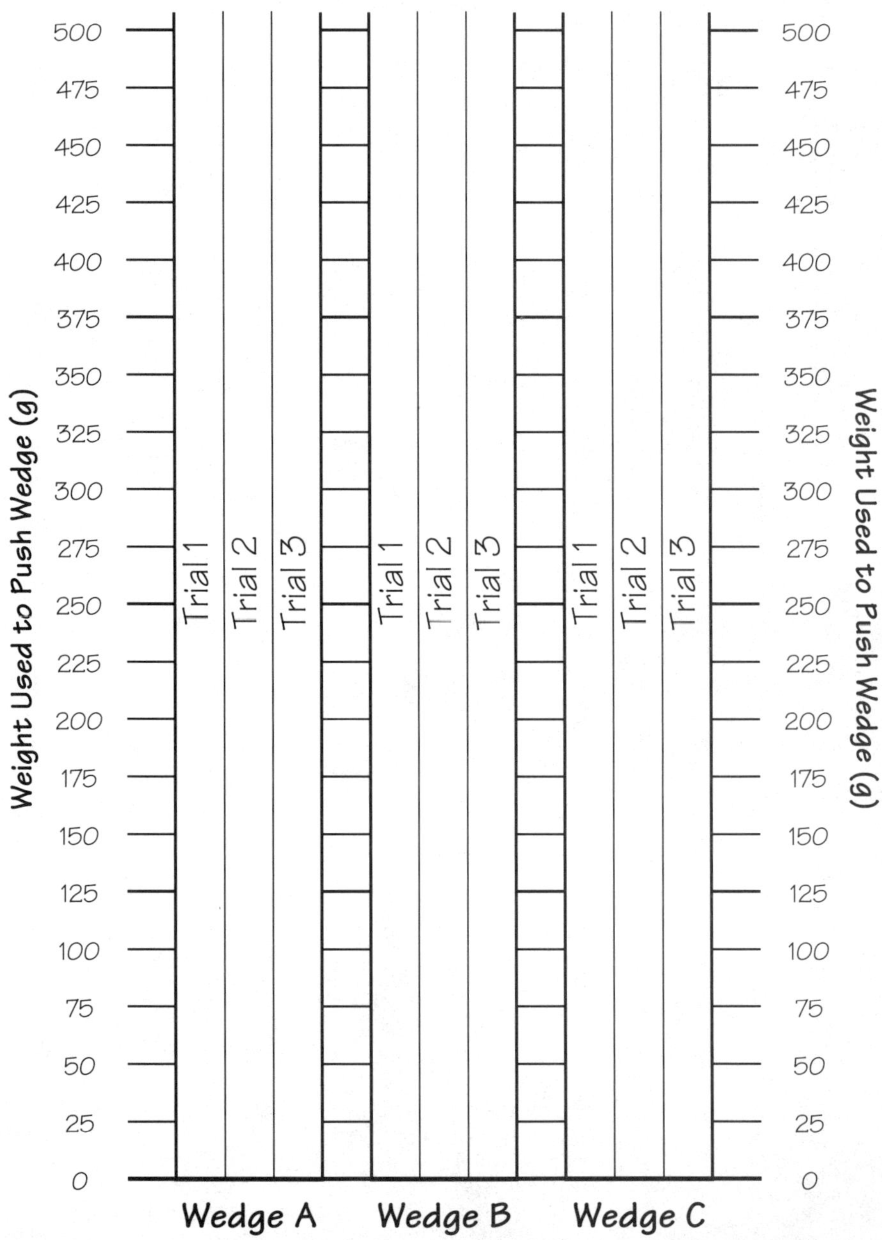

WEDGE-EASE

How does the length of a wedge affect how easily it pushes through something?

Rank the wedges in order from easiest to hardest to push through the books. Measure and record the length of each wedge. Gather the class data and find the class average. Use this data to make a line graph.

Class Data

		Put wedges in order from: Easiest	to	Hardest
	Wedge			
	Length (cm)			
Average weight used to push wedge (g)	Group 1			
	Group 2			
	Group 3			
	Group 4			
	Group 5			
	Group 6			
	Group 7			
	Group 8			
	Total			
	Average			

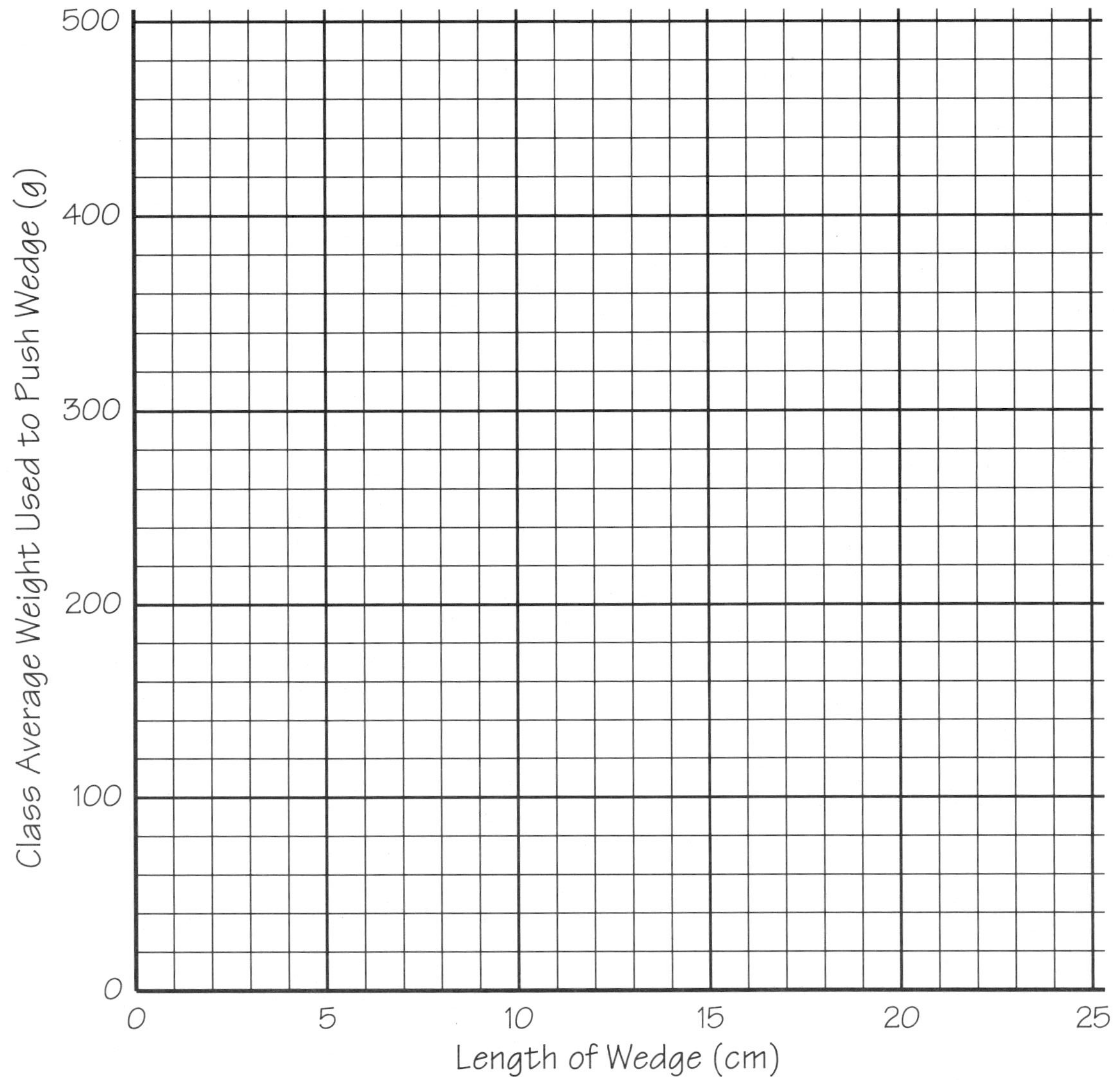

Nuts and Bolts

Topic Area
Simple machines: screws

Introductory Statement
Students will determine the mechanical advantage for different nuts and bolts by measuring how far the nut moves with each rotation.

Math
Using patterns and functions
Measuring
  length
Writing equations
Graphing

Science
Physical science
  Simple machines

Math/Science Process Skills
Observing
Collecting and recording data
Interpreting data
Applying and generalizing
Predicting
Inferring

Materials
Per group of two students:
1 3"-long, 1/2" diameter (12.75 mm) carriage bolt with nut
1 3"-long, 3/8" diameter (9.56 mm) carriage bolt with nut
ruler–calibrated in millimeters

Key Questions
1. How could a nut and bolt be used to measure small distances?
2. How could you use the distance a nut moves to determine the ideal mechanical advantage of the nut and bolt?

Background Information
A screw is a simple machine. It is an inclined plane wrapped in a spiral. The screw provides mechanical advantage by moving the resistance up the spiral a small distance while rotating the resistance around the spiral a long distance. The ideal mechanical advantage can be calculated by dividing the distance around the spiral by the distance the resistance moves up the spiral with each rotation.

The auger is an example of a screw. It lifts objects by rotating the shaft in the center of the auger's spiraling inclined plane. A load goes all the way around in the auger with each rotation but only moves up a small distance. The auger provides a means of moving things, but, because of friction, it is far from efficient.

Screws and bolts with nuts are used to hold things tightly together. Their high mechanical advantages provide the means to pull together objects with high resistances and to be held much tighter than they could without the bolt or screw.

Screws are also used in applications requiring precise measurement. The screw provides fine lateral movement when gross rotational movement is applied. The small precise movement allows screws to be used for calipers or for the adjustment of a car's alignment. The change from rotational to lateral motion provides a method for converting the spinning of an electrical motor to a precise lateral movement. This application allows machines to make strong, precise lateral motions.

The precision of screws provides the opportunity for students to gather real data with very consistent results. This data gives a straight line when graphed and excellent opportunities to discuss the relationship of data, graphs, and equations. The slope of the graphed line is the average movement per rotation. The Y-intercept is the original position of the nut on the bolt.

Management
1. This activity may require two 35-40 minute periods: one to gather data, one to graph and analyze data.
2. This activity is best done with partners. One person rotates the nut while the other person measures and records the data.
3. Before the activity, the teacher needs to mark the bolts and nuts with a permanent marker. Place a line on the bottom of each bolt's head (facing screw) from the shaft out to the edge. Place a line on one of the six sides of the nut from top to bottom. These lines will serve as alignment markers.
4. Before the activity, make sure all students can accurately read linear measurements to the nearest millimeter.
5. The number of nuts and bolts required for the class can be reduced in half if you have two groups share their bolts. One group can use the larger bolt while the other group uses the smaller. When the data has been collected, groups can switch bolts.

Procedure
1. Discuss the *Key Questions.*
2. Have students work in pairs. Distribute materials and the *Bolting Around* handout.
3. Direct students to wind the nut as far as it can be easily tightened with their fingers. Have them then back it off so that the lines that were drawn on the bolt and nut are aligned.
4. Have students measure to the nearest millimeter the distance from the bottom of the head of the bolt to the

closest part of the nut. This measurement should be recorded on the chart as the separation with 0 rotations.
5. Instruct students to rotate the nut back five rotations and align the marks.
6. Have them measure the separation and record it.
7. Repeat steps five and six until the chart is completed by making a total of 40 rotations.
8. Have students calculate the average change in separation per rotation, the circumference of the bolt, and the mechanical advantage of the nut and bolt system.
9. When the data for one bolt is completed, have students follow the same procedure for the other bolt.
10. Instruct students to make a line graph, making a line for each bolt on the *Going Nuts* handout.
11. Complete the page by having students write an equation for each bolt from the data and graph.

Discussion

1. What pattern do you see in the separation for five rotations on the chart? [It goes up about the same amount each five rotations.]
2. How could you use this pattern to predict what the separation would be for 50 rotations?
3. Which nut and bolt system separated faster?
4. Which nut and bolt system has the greater mechanical advantage?
5. What patterns do you see on the graph? [straight lines, sloping diagonally upwards toward the right showing a direct relationship between the number of rotations and the distance of separation]
6. How do the patterns on the graph relate to the nut and bolt? [Straight line means consistently the same amount of increase in separation per rotation. Upward diagonally means greater separation with each turn]
7. What differences do you see in the lines on the graph?
8. What do the differences on the graph tell you about the nuts and bolts systems. [The greater slope of the 12.75 mm bolt shows it is separating faster. The position at 0 rotations is the starting separation.]
9. How does the average change in separation per rotation show up on the graph? [Slope of the line–for every rotation the graphed line crosses, it also rises the average change]
10. How can you use the graph to predict what the separation will be with 50 rotations? [Extrapolate (extend the line) and find the separation for 50 rotations from the value on the y-axis of the graph]
11. How do the numbers in the equations relate to the lines on the graph and data? [The factor tells the slope of the line and is the average change. The addend tells the starting separation and is where the graphed line intercepts the y-axis.]
12. How can you use the equation to determine the separation at 50 rotations? [substitute 50 rotations and solve]
13. Does the distance of separation that you calculated for 50 rotations agree with the answer you found when you used the graph and extended the line to include 50 rotations? Explain.

Extensions

1. Use the nuts and bolts to measure the thickness of different objects and then check to see if they are accurate.
2. Bring in a vise and have students develop an equation to use for measuring. Have students explore the force generated by the vise.
3. Have students observe their world to discover as many different applications of screws.

Major Conceptual Components

- Energy is used to accomplish work.
- Work is done when something is moved.
- The amount of work done is determined by how much force was used to move the object and how far the object was moved.
- Mechanical advantage is a comparison of force produced by a machine compared to the force applied to the machine.
- Energy comes in many forms.
- Mechanical energy is due to the position or the movement of something.
- A machine is a device for multiplying force or changing the direction of force.
- Inclined planes can be used to raise objects with less force than a straight lift.
- A screw is an inclined plane that is wrapped around a shaft.

Nuts and Bolts

Bolting Around

Bolt Diameter: 9.56 mm

Nut Rotations	Separation (mm)
0	
5	
10	
15	
20	
25	
30	
35	
40	

Total change in separation for 40 rotations. [ ]

Average change in separation per rotation. [ ]

Circumference around the nut. [ ]

Mechanical Advantage (MA)

[ ] ÷ [ ] = [ ]

Circumference ÷ Average Change = MA

Bolt Diameter: 12.75 mm

Nut Rotations	Separation (mm)
0	
5	
10	
15	
20	
25	
30	
35	
40	

Total change in separation for 40 rotations. [ ]

Average change in separation per rotation. [ ]

Circumference around the nut. [ ]

Mechanical Advantage (MA)

[ ] ÷ [ ] = [ ]

Circumference ÷ Average Change = MA

Nuts and Bolts

Going Nuts

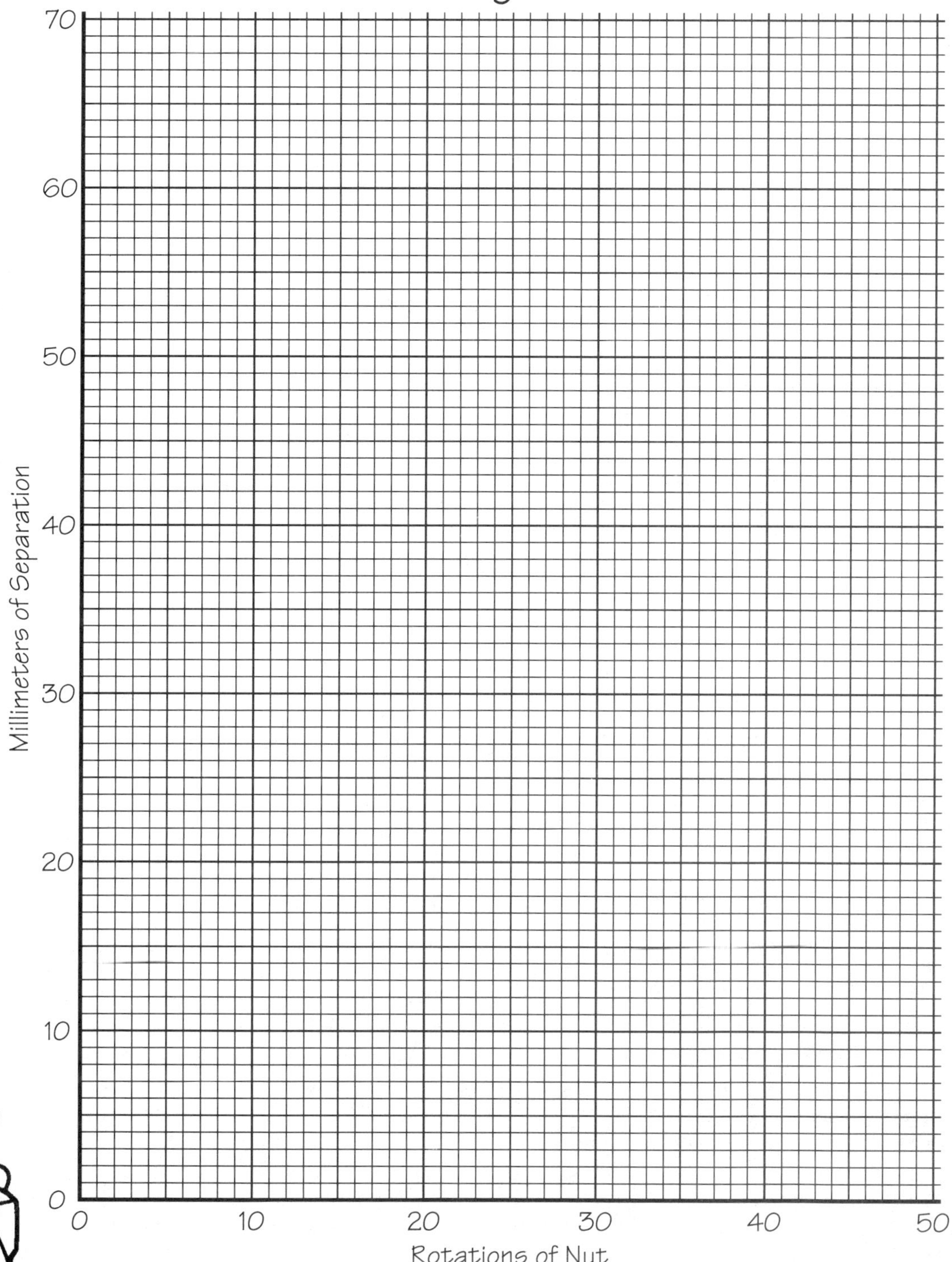

Can you write an equation for each bolt that will predict the separation (S) if you know the number of rotations (R)?

9.56 mm Bolt: S = (________ x R) + ________

12.75 mm Bolt: S = (x R) + ________

Watt Power!

Topic Area
Power

Introductory Statement
The students will determine their power in watts to see if there is a correlation between the weight of the person and the power they produce.

Math
Using computation
Using formulae
Measuring
- time
- distance
- weight

Counting
Graphing
- scattergrams

Correlating
Using calculators

Science
Physical science
- Simple machines

Math/Science Process Skills
Observing
Predicting
Collecting and recording data
Interpreting data
Inferring
Applying
Generalizing

Materials
Watch or clock with a second hand, or stopwatch
Measuring tape or ruler
Bathroom scale
Calculators

Key Questions
1. How many watts of power can you generate?
2. Who do you think can make the most power? Why?

Background Information
The rate of doing work is called *power* (P). It is calculated by dividing work by time (P = W/t) where work (W) is the amount of work done and time (t) is the time it took to get the work done.

A tractor and a student can both do the work of moving a pile of dirt. However, it is evident that the tractor is more powerful because it will get the work done in a shorter amount of time.

The metric unit of power is the watt (named after James Watt , the inventor of the steam engine). A watt is the power generated when a newton of force is moved one meter in one second.

When a student climbs onto a chair, power is generated. The power can be calculated by multiplying the student's weight (in newtons) by the height of the climb (in meters) and dividing the product by the number of seconds it takes to make the climb. Climbing onto a chair takes such a short time that timing a series of climbs provides greater accuracy.

If two people climb onto a chair at the same rate, the one weighing the most will have done the most work. As a result, students weighing more tend to generate more power than other lighter students.

Customarily in the United States, engines are rated in horsepower and electrical motors and appliances in watts. If students are interested in their horsepower rating, they can use the conversion rate of 750 watts equals 1 horsepower.

Management
1. This activity should take two 35-40 periods. One period will be occupied taking measurements and doing calculations, the second for sharing data and graphing.
2. Before doing this activity, the teacher must select a means for students to lift themselves. The activity is written assuming students will climb up onto a chair's seat. However, a long stairway may be used if it is available. The height of the stairway will need to be determined. Students will then need to be timed individually as they climb the stairs. The calculation chart will need to be modified.
3. This activity is best done in pairs. One person holds the chair and counts the other person's lifts.
4. Before doing this activity, the teacher must remind the students of safety. Partners must be sure to hold the chairs securely so they do not slip out from under the other students. The students climbing onto the chair must be careful not to slip.
5. For accuracy, students must make sure they lift their entire body the height of the chair's seat. To maintain this standard, have students stand straight on the floor with both feet. Their partners will count a lift each time they are standing straight with both feet on the seat of the chair.
6. The activity of climbing up and down on the chair may be done in unison. The teacher says "go" and times 30 seconds before telling the students to stop. The partners switch roles and 30 seconds is timed again.
7. If a metric scale is not available, have students divide their weight in pounds by 2.2 to obtain the kilogram equivalent.

Procedure
1. This activity requires students to use their weights in determining the power they generate. If some of your

students are sensitive about having their weights exposed to the rest of the class, procedures should be altered to accommodate those feelings.
2. Discuss with students the concept of power and the meaning of watts. You may want to display different items such as light bulbs and hair dryers that have the wattages listed on them.
3. Discuss *Key Questions*.
4. Have students get together with a partner. Students will record data on both themselves and their partner.
5. Ask students to predict how many watts they think they will be able to generate and record this prediction on the activity sheet.
6. Have students weigh themselves and record results in kilograms.
7. Direct students to convert their mass to newtons and record it.
8. Have students measure the height of the seat of the chair in meters and record it.
9. With one student holding the back of the chair to steady it, direct the other student to climb up and down on it for 30 seconds. The partner holding the chair will count the number of lifts and record it. Switch responsibilities and repeat the exercise. Record results.
10. Have students do the calculations to complete the record sheet and check their results with their partner's to see that both are the same.
11. Compile a class record by having students share their results.
12. Direct students to use the class data to make a scattergram.

Discussion

1. How did your predicted power compare to your actual power?
2. How well did you predict the most powerful person in your class?
3. What generalization can you make about a person's force and the power generated? [This can vary depending on the class data, but generally a person with greater force (weight) generates more horsepower.]
4. How can you validate this conclusion from the graph? [as the force goes up, the watts goes up]
5. Explain why someone with greater force would be able to generate more power? [They move more mass, so they do more work every time they move.]
6. Is there anyone in your class who is an exception to this generalization? Why might they fall outside the norm? [Often there is a very quick and agile student that makes significantly more lifts than the other students. You may not want to ask this question if you find the exceptions are overweight students that are significantly slowed by their weight.]
7. If you had to use your power to operate an electric appliance, what could you run with your power?

Extensions

1. Have students repeat the activity with a loaded backpack and see how it affects the watts they generate.
2. Have students do as many pull-ups as they can and time how long it takes to do them. Have them calculate the watts they can generate this way.
3. Have students go home and make a listing of appliances and the watts they require. Have students determine how many class members it would take to power each of the appliances.

Major Conceptual Components

- Energy is used to accomplish work.
- Work is done when something is moved.
- The amount of work done is determined by how much force was used to move the object and how far the object was moved.
- Power is the rate at which work is done.

WATT POWER

Individual Data

Student Name	Predicted Power (watts)	Mass of Student (kg)	Force kg x 9.8 (N)	Height of Lift (m)	Number of Lifts in 30 Seconds	Newton-Meters (N x m x Lifts)	Actual Power newton • Meter = Watts 30 Seconds

Class Data

Student Name	Force	Watts
1.		
2.		
3.		
4.		
5.		
6.		
7.		
8.		
9.		
10.		
11.		
12.		
13.		
14.		
15.		
16.		
17.		

Student Name	Force	Watts
18.		
19.		
20.		
21.		
22.		
23.		
24.		
25.		
26.		
27.		
28.		
29.		
30.		
31.		
32.		
Total		
Average		

Watt Power Graph

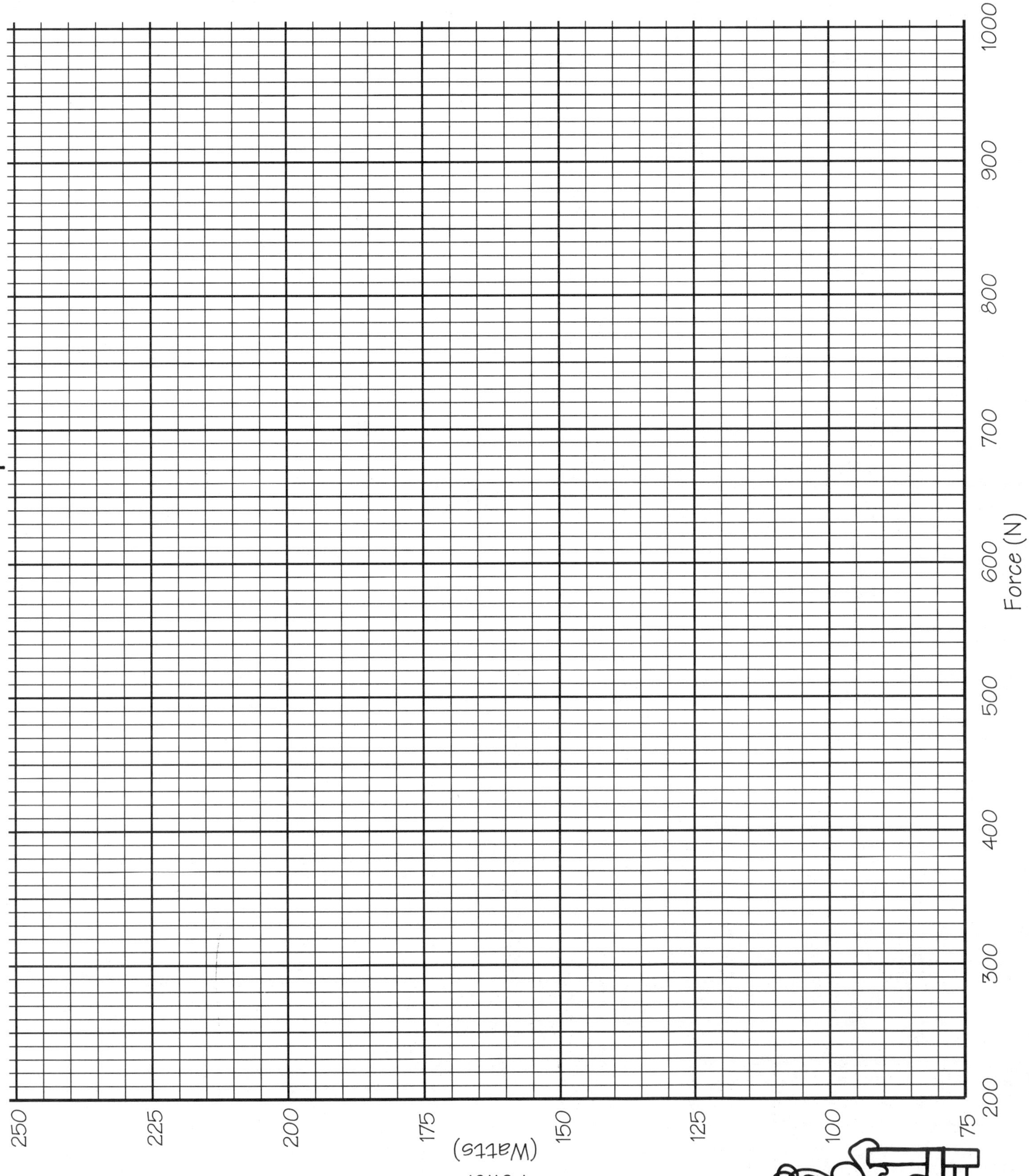

The AIMS Program

AIMS is the acronym for "**A**ctivities **I**ntegrating **M**athematics and **S**cience." Such integration enriches learning and makes it meaningful and holistic. AIMS began as a project of Fresno Pacific University to integrate the study of mathematics and science in grades K-9, but has since expanded to include language arts, social studies, and other disciplines.

AIMS is a continuing program of the non-profit AIMS Education Foundation. It had its inception in a National Science Foundation funded program whose purpose was to explore the effectiveness of integrating mathematics and science. The project directors in cooperation with 80 elementary classroom teachers devoted two years to a thorough field-testing of the results and implications of integration.

The approach met with such positive results that the decision was made to launch a program to create instructional materials incorporating this concept. Despite the fact that thoughtful educators have long recommended an integrative approach, very little appropriate material was available in 1981 when the project began. A series of writing projects have ensued and today the AIMS Education Foundation is committed to continue the creation of new integrated activities on a permanent basis.

The AIMS program is funded through the sale of this developing series of books and proceeds from the Foundation's endowment. All net income from program and products flows into a trust fund administered by the AIMS Education Foundation. Use of these funds is restricted to support of research, development, and publication of new materials. Writers donate all their rights to the Foundation to support its on-going program. No royalties are paid to the writers.

The rationale for integration lies in the fact that science, mathematics, language arts, social studies, etc., are integrally interwoven in the real world from which it follows that they should be similarly treated in the classroom where we are preparing students to live in that world. Teachers who use the AIMS program give enthusiastic endorsement to the effectiveness of this approach.

Science encompasses the art of questioning, investigating, hypothesizing, discovering, and communicating. Mathematics is the language that provides clarity, objectivity, and understanding. The language arts provide us powerful tools of communication. Many of the major contemporary societal issues stem from advancements in science and must be studied in the context of the social sciences. Therefore, it is timely that all of us take seriously a more holistic mode of educating our students. This goal motivates all who are associated with the AIMS Program. We invite you to join us in this effort.

Meaningful integration of knowledge is a major recommendation coming from the nation's professional science and mathematics associations. The American Association for the Advancement of Science in *Science for All Americans* strongly recommends the integration of mathematics, science, and technology. The National Council of Teachers of Mathematics places strong emphasis on applications of mathematics such as are found in science investigations. AIMS is fully aligned with these recommendations.

Extensive field testing of AIMS investigations confirms these beneficial results.

1. Mathematics becomes more meaningful, hence more useful, when it is applied to situations that interest students.
2. The extent to which science is studied and understood is increased, with a significant economy of time, when mathematics and science are integrated.
3. There is improved quality of learning and retention, supporting the thesis that learning which is meaningful and relevant is more effective.
4. Motivation and involvement are increased dramatically as students investigate real-world situations and participate actively in the process.

We invite you to become part of this classroom teacher movement by using an integrated approach to learning and sharing any suggestions you may have. The AIMS Program welcomes you!

AIMS Education Foundation Programs

A Day with AIMS

Intensive one-day workshops are offered to introduce educators to the philosophy and rationale of AIMS. Participants will discuss the methodology of AIMS and the strategies by which AIMS principles may be incorporated into curriculum. Each participant will take part in a variety of hands-on AIMS investigations to gain an understanding of such aspects as the scientific/mathematical content, classroom management, and connections with other curricular areas. *A Day with AIMS* workshops may be offered anywhere in the United States. Necessary supplies and take-home materials are usually included in the enrollment fee.

A Week with AIMS

Throughout the nation, AIMS offers many one-week workshops each year, usually in the summer. Each workshop lasts five days and includes at least 30 hours of AIMS hands-on instruction. Participants are grouped according to the grade level(s) in which they are interested. Instructors are members of the AIMS Instructional Leadership Network. Supplies for the activities and a generous supply of take-home materials are included in the enrollment fee. Sites are selected on the basis of applications submitted by educational organizations. If chosen to host a workshop, the host agency agrees to provide specified facilities and cooperate in the promotion of the workshop. The AIMS Education Foundation supplies workshop materials as well as the travel, housing, and meals for instructors.

AIMS One-Week Perspectives Workshops

Each summer, Fresno Pacific University offers AIMS one-week workshops on its campus in Fresno, California. AIMS Program Directors and highly qualified members of the AIMS National Leadership Network serve as instructors.

The Science Festival and the Festival of Mathematics

Each summer, Fresno Pacific University offers a Science Festival and a Festival of Mathematics. These festivals have gained national recognition as inspiring and challenging experiences, giving unique opportunities to experience hands-on mathematics and science in topical and grade-level groups. Guest faculty includes some of the nation's most highly regarded mathematics and science educators. Supplies and take-home materials are included in the enrollment fee.

The AIMS Instructional Leadership Program

This is an AIMS staff-development program seeking to prepare facilitators for leadership roles in science/math education in their home districts or regions. Upon successful completion of the program, trained facilitators become members of the AIMS Instructional Leadership Network, qualified to conduct AIMS workshops, teach AIMS in-service courses for college credit, and serve as AIMS consultants. Intensive training is provided in mathematics, science, process and thinking skills, workshop management, and other relevant topics.

College Credit and Grants

Those who participate in workshops may often qualify for college credit. If the workshop takes place on the campus of Fresno Pacific University, that institution may grant appropriate credit. If the workshop takes place off-campus, arrangements can sometimes be made for credit to be granted by another college or university. In addition, the applicant's home school district is often willing to grant in-service or professional development credit. Many educators who participate in AIMS workshops are recipients of various types of educational grants, either local or national. Nationally known foundations and funding agencies have long recognized the value of AIMS mathematics and science workshops to educators. The AIMS Education Foundation encourages educators interested in attending or hosting workshops to explore the possibilities suggested above. Although the Foundation strongly supports such interest, it reminds applicants that they have the primary responsibility for fulfilling *current* requirements.

For current information regarding the programs described above, please complete the following:

Information Request

Please send current information on the items checked:

___ *Basic Information Packet* on AIMS materials
___ *Festival of Mathematics*
___ *Science Festival*
___ *AIMS Instructional Leadership Program*
___ *AIMS One-Week Perspectives* workshops
___ *A Week with AIMS* workshops
___ Hosting information for *A Day with AIMS* workshops
___ Hosting information for *A Week with AIMS* workshops

Name ______________________ Phone ______________________

Address ____

Street City State Zip

We invite you to subscribe to AIMS!

Each issue of AIMS contains a variety of material useful to educators at all grade levels. Feature articles of lasting value deal with topics such as mathematical or science concepts, curriculum, assessment, the teaching of process skills, and historical background. Several of the latest AIMS math/science investigations are always included, along with their reproducible activity sheets. As needs direct and space allows, various issues contain news of current developments, such as workshop schedules, activities of the AIMS Instructional Leadership Network, and announcements of upcoming publications.

AIMS is published monthly, August through May. Subscriptions are on an annual basis only. A subscription entered at any time will begin with the next issue, but will also include the previous issues of that volume. Readers have preferred this arrangement because articles and activities within an annual volume are often interrelated.

Please note that an AIMS subscription automatically includes duplication rights for one school site for all issues included in the subscription. Many schools build cost-effective library resources with their subscriptions.

YES! I am interested in subscribing to AIMS.

Name ______________________ Home Phone ______________

Address ______________________ City, State, Zip ______________

Please send the following volumes (subject to availability):

_____ Volume V	(1990-91)	$30.00		_____ Volume X	(1995-96)	$30.00
_____ Volume VI	(1991-92)	$30.00		_____ Volume XI	(1996-97)	$30.00
_____ Volume VII	(1992-93)	$30.00		_____ Volume XII	(1997-98)	$30.00
_____ Volume IX	(1994-95)	$30.00		_____ Volume XIII	(1998-99)	$30.00

_____ **Limited offer: Volumes XIII & XIV (1998-2000) $55.00**
(Note: Prices may change without notice)

Check your method of payment:

❒ Check enclosed in the amount of $ __________

❒ Purchase order attached (Please include the P.O.#, the authorizing signature, and position of the authorizing person.)

❒ Credit Card ❒ Visa ❒ MasterCard Amount $ ______________

Card # ______________________ Expiration Date ___________

Signature ______________________ Today's Date ___________

Make checks payable to **AIMS Education Foundation.**
Mail to AIMS magazine, P.O. Box 8120, Fresno, CA 93747-8120.
Phone (209) 255-4094 or (888) 733-2467 FAX (209) 255-6396
AIMS Homepage: http://www.AIMSedu.org/

AIMS Program Publications

GRADES K-4 SERIES

Bats Incredible
Brinca de Alegria Hacia la Primavera con las Matemáticas y Ciencias
Cáete de Gusto Hacia el Otoño con la Matemáticas y Ciencias
Cycles of Knowing and Growing
Fall Into Math and Science
Field Detectives
Glide Into Winter With Math and Science
Hardhatting in a Geo-World (Revised Edition, 1996)
Jaw Breakers and Heart Thumpers (Revised Edition, 1995)
Los Cincos Sentidos
Overhead and Underfoot (Revised Edition, 1994)
Patine al Invierno con Matemáticas y Ciencias
Popping With Power (Revised Edition, 1996)
Primariamente Física (Revised Edition, 1994)
Primarily Earth
Primariamente Plantas
Primarily Physics (Revised Edition, 1994)
Primarily Plants
Sense-able Science
Spring Into Math and Science
Under Construction

GRADES K-6 SERIES

Budding Botanist
Critters
El Botanista Principiante
Mostly Magnets
Ositos Nada Más
Primarily Bears
Principalmente Imanes
Water Precious Water

GRADES 5-9 SERIES

Actions with Fractions
Brick Layers
Conexiones Eléctricas
Down to Earth
Electrical Connections
Finding Your Bearings (Revised Edition, 1996)
Floaters and Sinkers (Revised Edition, 1995)
From Head to Toe
Fun With Foods
Historical Connections in Mathematics, Volume I
Historical Connections in Mathematics, Volume II
Historical Connections in Mathematics, Volume III
Machine Shop
Magnificent Microworld Adventures
Math + Science, A Solution
Off the Wall Science: A Poster Series Revisited
Our Wonderful World
Out of This World (Revised Edition, 1994)
Pieces and Patterns, A Patchwork in Math and Science
Piezas y Diseños, un Mosaic de Matemáticas y Ciencias
Soap Films and Bubbles
Spatial Visualization
The Sky's the Limit (Revised Edition, 1994)
The Amazing Circle, Volume 1
Through the Eyes of the Explorers:
Minds-on Math & Mapping
What's Next, Volume 1
What's Next, Volume 2
What's Next, Volume 3

AIMS Duplication Rights Program

AIMS has received many requests from school districts for the purchase of unlimited duplication rights to AIMS materials. In response, the AIMS Education Foundation has formulated the program outlined below. There is a built-in flexibility which, we trust, will provide for those who use AIMS materials extensively to purchase such rights for either individual activities or entire books.

It is the goal of the AIMS Education Foundation to make its materials and programs available at reasonable cost. All income from the sale of publications and duplication rights is used to support AIMS programs; hence, strict adherence to regulations governing duplication is essential. Duplication of AIMS materials beyond limits set by copyright laws and those specified below is strictly forbidden.

Limited Duplication Rights

Any purchaser of an AIMS book may make up to *200 copies* of any activity in that book for use at *one school site*. Beyond that, rights must be purchased according to the appropriate category.

Unlimited Duplication Rights for Single Activities

An individual or school may purchase the right to make an unlimited number of copies of a single activity. The royalty is $5.00 per activity per school site.

Examples: 3 activities x 1 site x $5.00 = $15.00
9 activities x 3 sites x $5.00 = $135.00

Unlimited Duplication Rights for Entire Books

A school or district may purchase the right to make an unlimited number of copies of a single, *specified* book. The royalty is $20.00 per book per school site. This is in addition to the cost of the book.

Examples: 5 books x 1 site x $20.00 = $100.00
12 books x 10 sites x $20.00 = $2400.00

Magazine/Newsletter Duplication Rights

Members of the AIMS Education Foundation who purchase the *AIMS* magazine/*Newsletter* are hereby granted permission to make up to 200 copies of any portion of it, provided these copies will be used for educational purposes.

Workshop Instructors' Duplication Rights

Workshop instructors may distribute to registered workshop participants a maximum of 100 copies of any article and/or 100 copies of no more than eight activities, provided these six conditions are met:

1. Since all AIMS activities are based upon the *AIMS Model of Mathematics* and the *AIMS Model of Learning*, leaders must include in their presentations an explanation of these two models.
2. Workshop instructors must relate the AIMS activities presented to these basic explanations of the AIMS philosophy of education.
3. The copyright notice must appear on all materials distributed.
4. Instructors must provide information enabling participants to apply for membership in the AIMS Education Foundation or order books from the Foundation.
5. Instructors must inform participants of their limited duplication rights as outlined below.
6. Only student pages may be duplicated.

Written permission must be obtained for duplication beyond the limits listed above. Additional royalty payments may be required.

Workshop Participants' Rights

Those enrolled in workshops in which AIMS student activity sheets are distributed may duplicate a maximum of 35 copies or enough to use the lessons one time with one class, whichever is less. Beyond that, rights must be purchased according to the appropriate category.

Application for Duplication Rights

The purchasing agency or individual must clearly specify the following:

1. Name, address, and telephone number
2. Titles of the books for Unlimited Duplication Rights contracts
3. Titles of activities for Unlimited Duplication Rights contracts
4. Names and addresses of school sites for which duplication rights are being purchased.

NOTE: Books to be duplicated must be purchased separately and are not included in the contract for Unlimited Duplication Rights.

The requested duplication rights are automatically authorized when proper payment is received, although a *Certificate of Duplication Rights* will be issued when the application is processed.

Address all correspondence to: **Contract Division**
AIMS Education Foundation
P.O. Box 8120
Fresno, CA 93747-8120